Traffic Control Devices 2018

Transportation Research Record: Journal of the Transportation Research Board

Volume 2672, Issue 21

ISBN: 978-1-5108-7826-6

Printed from e-media with permission by:

Curran Associates, Inc.
57 Morehouse Lane
Red Hook, NY 12571

Some format issues inherent in the e-media version may also appear in this print version.

Published by Sage Publications USA

ISBN (Print) 978-1-5108-7826-6
ISBN 2018 Printed Set (All Issues) 978-1-5108-7735-1

Printed by Curran Associates, Inc. (2019)

Additional copies of this publication are available from:

Curran Associates, Inc.
57 Morehouse Lane
Red Hook, NY 12571 USA
Phone: 845-758-0400
Fax: 845-758-2633
Email: curran@proceedings.com
Web: www.proceedings.com

Volume 2672 Number 21 December 2018

TRR
JOURNAL OF THE TRANSPORTATION RESEARCH BOARD

Contents

Traffic Control Devices 2018

Articles

Evaluation of Intersection Conflict Warning System: A Critical Gap Analysis 1
Raju Thapa, Shauna Hallmark, Neal Hawkins, and Skylar Knickerbocker

Safety Effects of Turning Movement Restrictions at Stop-Controlled Intersections 10
Thanh Q. Le, Frank Gross, and Tim Harmon

Safety Effects of Flashing Yellow Arrows Used in Protected Permitted Phasing: Comparison of Full Bayes and Empirical Bayes Results 20
Justice Appiah, F. Adam King, Michael D. Fontaine, and Benjamin H. Cottrell, Jr.

Safety Evaluation of Alternative Audible Lane Departure Warning Treatments in Reducing Traffic Crashes: An Empirical Bayes Observational Before–After Study 30
Lingtao Wu, Srinivas R. Geedipally, and Adam M. Pike

Closed-Course Study to Assess the Conspicuity of Wrong-Way Driving Countermeasures 41
Melisa D. Finley and Jeffrey D. Miles

Field Evaluation of the Dilemma Zone Protection System at Suburban Intersections 51
Sung Yoon Park, Chien-Lun Lan, Ranteg S. Rao, and Gang-Len Chang

EX OFFICIO MEMBERS

---

* Membership as of August 2018.

TRB

TRANSPORTATION RESEARCH BOARD

CORE SPONSORS

Transportation Departments of the 50 States and the District of Columbia

Federal Government

U.S. Department of Transportation
- Federal Aviation Administration (FAA)
- Federal Highway Administration
- Federal Motor Carrier Safety Administration (FMCSA)
- Federal Railroad Administration (FRA)
- Federal Transit Administration
- National Highway Traffic Safety Administration (NHTSA)
- Office of the Assistant Secretary of Transportation for Research and Technology (OST-R)

U.S. Air Force Civil Engineer Center (AFCEC)
U.S. Coast Guard (USCG)
U.S. Department of Energy
U.S. Department of the Interior (DOI)
U.S. Environmental Protection Agency (EPA)

Nongovernmental Organizations

American Public Transportation Association (APTA)
Association of American Railroads (AAR)
California Air Resources Board

The National Academies of

SCIENCES · ENGINEERING · MEDICINE

The **National Academy of Sciences** was established in 1863 by an Act of Congress, signed by President Lincoln, as a private, nongovernmental institution to advise the nation on issues related to science and technology. Members are elected by their peers for outstanding contributions to research. Dr. Marcia McNutt is president.

The **National Academy of Engineering** was established in 1964 under the charter of the National Academy of Sciences to bring the practices of engineering to advising the nation. Members are elected by their peers for extraordinary contributions to engineering. Dr. C. D. Mote, Jr., is president.

The **National Academy of Medicine** (formerly the Institute of Medicine) was established in 1970 under the charter of the National Academy of Sciences to advise the nation on medical and health issues. Members are elected by their peers for distinguished contributions to medicine and health. Dr. Victor J. Dzau is president.

The three Academies work together as the National Academies of Sciences, Engineering, and Medicine to provide independent, objective analysis and advice to the nation and conduct other activities to solve complex problems and inform public policy decisions. The Academies also encourage education and research, recognize outstanding contributions to knowledge, and increase public understanding in matters of science, engineering, and medicine.

Learn more about the National Academies of Sciences, Engineering, and Medicine at **www.national-academies.org.**

The **Transportation Research Board** is one of seven major programs of the National Academies of Sciences, Engineering, and Medicine. The mission of the Transportation Research Board is to increase the benefits that transportation contributes to society by providing leadership in transportation innovation and progress through research and information exchange, conducted within a setting that is objective, interdisciplinary, and multimodal. The Board's varied activities annually engage about 7,000 engineers, scientists, and other transportation researchers and practitioners from the public and private sectors and academia, all of whom contribute their expertise in the public interest. The program is supported by state transportation departments, federal agencies including the component administrations of the U.S. Department of Transportation, and other organizations and individuals interested in the development of transportation.

Learn more about the Transportation Research Board at **www.TRB.org**.

Article

TRR
JOURNAL OF THE TRANSPORTATION RESEARCH BOARD

Transportation Research Record
2018, Vol. 2672(21) 1–9

DOI: 10.1177/0361198118777357
journals.sagepub.com/home/trr
SAGE

Evaluation of Intersection Conflict Warning System: A Critical Gap Analysis

Raju Thapa[1], Shauna Hallmark[1], Neal Hawkins[1], and Skylar Knickerbocker[1]

Abstract

Rural intersections account for around 30% of crashes in rural areas and 6% of all fatal crashes. An Intersection Conflict Warning System (ICWS) is a unique solution to address rural intersection safety. ICWS are typically installed at the minor approaches to two-way stop-controlled intersections to reduce the number of fatalities. Studies indicate these systems result in lower intersection approach speeds, reduced conflicts, and improved driver-gap selection. However, some sites have experienced increases in the number of minor crashes. Although there are positive ICWS examples, their overall effectiveness is not well established. The objective of this research was to evaluate driving behavior at stop-controlled approaches with and without ICWS and to evaluate the spillover effect of ICWS on other adjacent control intersections where the treatment has not been applied. The study examined behavior at five intersections (treatment sites) in Minnesota where an ICWS was installed. For comparison, an additional five similar (control sites) were identified in proximity to each treatment intersection. Data were collected using a video camera array at three different time frames: before, one month, and 12 months after the installation of system. The data were analyzed using a critical gap approach. The analysis shows that the ICWS improved driver gap acceptance at the treatment sites, only at the 12-month period, and that there was no "spillover effect" at the adjacent control sites. When gap acceptance was further compared by type of stop, critical gap selection was shown to increase for drivers making both complete and rolling stops.

Rural intersections account for around 30% of crashes in rural areas of which 6% are fatal crashes. Crashes at rural stop-controlled intersections tend to be more severe due to higher speeds along the uncontrolled major approach and longer emergency response times (*1*). Higher speeds also create issues for drivers on minor stop-controlled approaches in judging and selecting acceptable gaps. Coupled with sight distance issues in some situations, crash likelihood is increased.

Effectiveness of ICWS

Intersection Conflict Warning Systems (ICWS) have been used at rural intersections to warn drivers on the minor approach of on-coming traffic and to warn drivers on the main approach of vehicles entering from the minor approach (*2*). Early studies have indicated that ICWS can result in lower intersection approach speeds, reduced conflicts, improved compliance with traffic control, and improved gap selection (*3–6*). Weidemann et al. (*4*) evaluated an intersection in Minnesota which had been instrumented with an ICWS. They found a 4.5% decrease in speed when the system was active compared with before installation. Rakauskas et al. (*5*) also studied ICWS in Minnesota and reported an increase in 80th percentile gap size for through vehicles when the system was activated. However, there has been some evidence that when the ICWS is not activated, drivers are less likely to comply with the stop sign and some sites have experienced minor crash increases (*4*). Weidemann et al. (*4*) found an increase in crashes (13%–24%).

An early study on ICWS evaluated a single intersection where an ICWS had been installed (*3*). Speed was collected before installation and at four months after to evaluate the impact of the system. Mean speeds decreased by 1.4 mph at the intersection and by 0.5 mph at the point drivers observed the sign.

Simpson and Troy (*7*) evaluated the effectiveness of ICWS in North Carolina. Several different configurations of ICWS are utilized throughout the state. The

[1]Institute for Transportation at Iowa State University, Ames, IA

Corresponding Author:
Address correspondence to Shauna Hallmark: shallmar@iastate.edu

study included 74 intersections. Since the systems were installed at different times, the before and after period differed but the average before period was five years and at least one year of after data were available for each site. Control intersections with similar geometric characteristics were selected and a before and after analysis was conducted using empirical Bayes. CMFs were developed based on different configurations and geometry, however in general they reported a crash modification factor for total crashes as 0.75. Kwon and Ismail (*8*) evaluated an early version of the current ICWS system in Minnesota. The system detects vehicles at both the minor and major approaches and activates LED blinker warning signs, "Vehicle Approaching When Flashing." The number of roll-through stops observed decreased from 28% before to 14% after installation. They also compared roll-through stops while the system was activated and found only 1% of vehicles engaged in a roll-through while the system was active. Speeds were also compared with the average speed during peak period decreasing from 52.0 to 50.8 mph. When the system was active, the average speed during the peak period was 47.7 mph.

Himes et al. (*9*) conducted an empirical Bayes before–after analysis using intersections from Minnesota, Missouri, and North Carolina to evaluate the effectiveness of ICWS. They developed crash modification factors for total crashes of 0.73 and 0.83 for two-lane roads intersecting with a two-lane side road and for four-lane roads intersecting at a two-lane minor road respectively. Manser and Creaser (*10*) used a driving simulator to evaluate the effectiveness of Cooperative Intersection Collision Avoidance System – Stop Sign Assist (CICAS-SSA) signs. A total of 21 participants were examined in a driving simulator both with and without the system being active. The result showed increases in the rejected gaps closer to the system activation threshold (*10*).

Objectives

Although ICWS show promise, their effectiveness has not been well established since robust crash analyses are not yet available due to short post-treatment, or after, periods. In a few cases, results suggested the system may decrease minor street stop sign compliance when drivers were not presented with a warning due to the absence of on-coming traffic. There was concern that if true, this may lead to a disregard for stop signs at adjacent intersections or in general.

The Minnesota Department of Transportation (MnDOT) has installed several ICWS in the last several years due to the perceived safety benefits. However, there were also concerns that negative changes in driver behavior at treatment intersections may result in a spillover effect at surrounding intersections where the system was not present. For instance, some evidence suggested drivers are more likely to come to a full stop when the system is active but less likely when the system is not active. This type of conditioning could lead to a general disregard of stop signs. Additionally, since the systems are reasonably expensive, agencies are interested in more evidence of their effectiveness. ICWS in Minnesota were evaluated at both treatment and adjacent control sites to assess the impacts on driver behavior including gap selection which was one of the variables evaluated since there is some correlation to crash likelihood.

Data Collection

Site Selection

The following summarizes the selection of treatment and control sites. Treatment sites are those intersections where an ICWS was installed. Control sites are intersections near the treatment sites which are expected to have similar drivers and road characteristics but do not have an ICWS. It should be noted that control sites are not true control sites in the traditional sense for safety studies. Typically control sites are selected to represent base conditions and reflect changes in crash or speed patterns related to characteristics other than an installed countermeasure. In this study, control intersections were selected to assess spillover effects. The intention was to determine whether drivers in a particular area acclimated to the ICWS technology with a corresponding change in behavior overall. The term is used to differentiate treatment from non-treatment sites to be consistent within the study. Figure 1 below shows an example of ICWS installed at a treatment site.

A list of all known sites where ICWS were planned for installation in Minnesota during 2014 was provided by MnDOT. Each potential test site was examined to determine its suitability for data collection which included

Figure 1. Example of ICWS.

consideration of trees/shrubs, steep ditches, or other objects along the roadway which would make it difficult to set up the video equipment.

Treatment sites were also assessed for uncommon characteristics which would make it difficult to select a similar control site. These unique features could include a railroad, significant vertical or horizontal curvature along one approach near the intersection, sight distance issues, and so forth. Due to a scheduling delay in installing the ICWS, potential treatment sites were further reduced to those which were likely to be installed within the analysis time period.

At least one control site was selected for each potential treatment intersection. Control sites were selected based on having similar roadway geometry, traffic volumes, and traffic control as the nearby treatment site. For instance, since all of the test intersections have two-way stop control, only intersections with a two-way stop were considered for the control intersections. Other characteristics were matched as much as possible such as selecting similar mainline roadway characteristics. Consequently, a control site on a divided highway was selected if the treatment site was on a divided highway. Other characteristics for selection included turning lane configuration and intersection angle.

When possible, the control intersection was selected along the same minor roadway as the treatment site. As a result, similar or even the same drivers would be collected at both intersections. The distance between the test intersection and control intersection was also an important consideration. Control intersections that were between one and five miles away from the treatment intersections were given first priority. The volume of the major and minor roadways was also a deciding factor. An effort was made to have major and minor roadway volumes that were similar between the treatment and control intersections. It was also necessary to have a large enough volume so that the system would be activated for a reasonable amount of time. Finally, the ability to situate data collection equipment along the control intersection was considered. Treatment and control sites are shown in Figure 2.

Data Collection

A data collection and reduction plan was developed before collection of data in the field. Video and volume data were collected in the field over three different time periods using a video camera array in which different cameras were positioned at the intersection to record the overall movement of vehicles along both the major and minor streets. Data were collected before the installation of the system, one month after the installation of system and 12 months after installation. Data were collected at

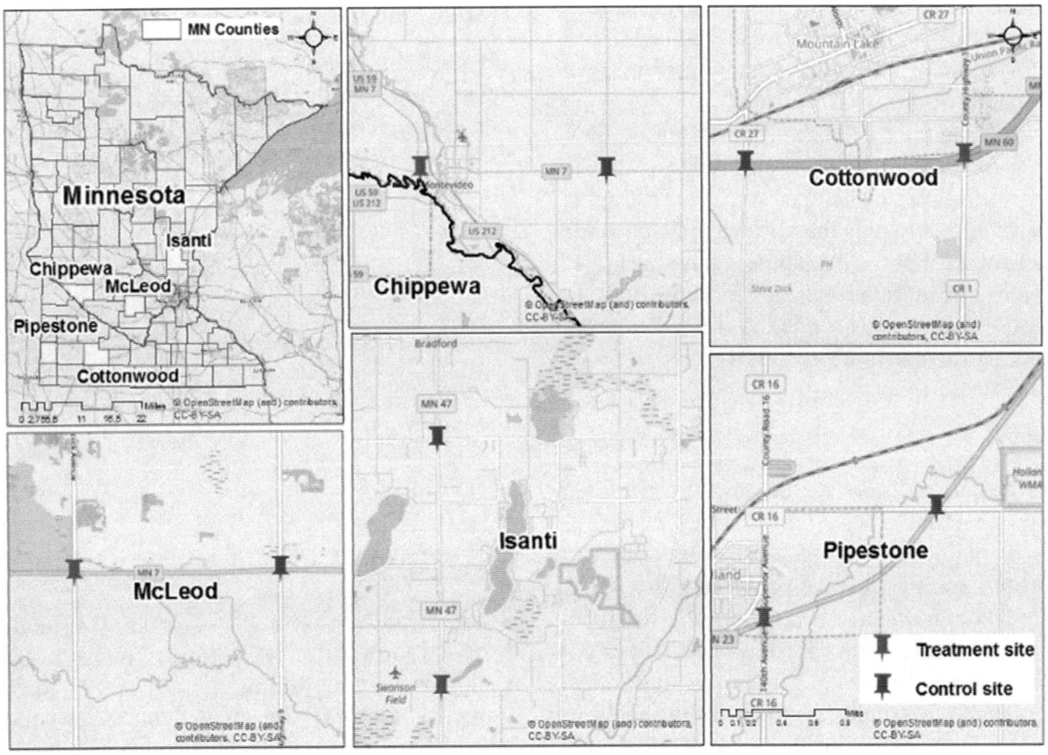

Figure 2. Treatment and control sites in Minnesota.

the same time for each treatment/control pair of intersections. Data were collected for approximately seven days.

Data Reduction

Given the time required for data reduction from video, a sampling plan was developed where a random time generator was used to generate a start time for each hour. Based on that start time, the first five (minor approach) vehicles in free flow condition were then reduced. Videos were reduced only during weekdays and during the hours of 6 a.m. to 8 p.m. since the cameras did not have infrared capabilities. Data were collected at the same time for the treatment and control intersections with the same random timeframe used for each intersection pair.

Gaps were reduced for minor approach vehicles. The first gap for a particular vehicle was reduced when the minor stream vehicle arrived at the intersection and each successive gap for that vehicle was reduced until the vehicle accepted a gap. A gap was defined as the time headway between front bumpers of two successive main stream vehicles. Additional coding features included type of gap (accepted or rejected), type of vehicle turning maneuver, time of a day, whether the ICWS was activated (system activation threshold of 6.5 seconds), weather condition, time spent at the minor approach, and type of stop. Vehicles were categorized as Group 1 or Group 2. Group 1 consisted of passenger vehicles such as pick-up trucks, mini-vans, regular passenger vehicles, and motorcycles. Group 2 consisted of large vehicles such as trucks, buses, and farm vehicles. Driver stopping behavior was coded as complete stop, rolling stop, or non-stop. Type of turn was coded as right, left, or through.

All gaps were reduced by the data reductionists, however, in many cases there were large major roadway vehicle gaps due to significant spacing between vehicles. The significantly large gaps did not require the same gap identification and selection process. Several researchers have identified this threshold as 12 or more seconds and indicated that for practical purposes there was no need to differentiate between gap sizes greater than 12 seconds (*11*, *12*).

Methodology

Evaluation of the critical gap is a common metric used to define gap acceptance behavior at intersections. A critical gap is the gap size at which a driver would reject a smaller gap and accept a gap of equal or greater size. Several different methods have been proposed to evaluate critical gaps at unsignalized intersections (*13*). Patil and Pawar, 2014 estimated temporal and spatial critical gaps at four-legged, partially controlled intersections using Raff's, Logit, Lag, Ashworth's, and Maximum Likelihood methods (*14*). Researchers also studied gap acceptance behavior at atypical stop-controlled intersections using different critical gap size estimation methods like Siegloch, Greenshields, Raff's, Acceptance curve and Logit methods (*15*).

Another study analyzed critical gap size using Raff's, Greenshield, Acceptance curve, Maximum Likelihood, Logit and Probit methods to check gap acceptance behavior in mandatory lane change locations under congested and uncongested traffic conditions (*16*). In summary, critical gaps can be analyzed using either a deterministic or probabilistic approach. Raff's method and Logit model methods were selected from Deterministic and Probabilistic approaches, respectively, for this study.

Deterministic Approach

Raff's method defines the critical gap as the size of gap for which the number of accepted gaps shorter than it is equal to the number of rejected gaps longer than it (*17*). Cumulative frequencies of accepted and rejected gap are plotted against the gap size and the intersection between the two curves is defined as the critical gap. One criticism of this approach is the inability to handle missing accepted or rejected gap information (*13*). To solve this problem, either lag or gap data can be merged assuming no statistical difference between lag and gap data (*18*) or by analyzing "lag only" and "gap only". For this study, critical gap was analyzed assuming no significant difference between lag and gap data.

Probabilistic Approach

Fitzpatrick (1991) used logit models to calculate the critical gap (*18*) which treats accepting or rejecting a gap as a dependent variable. The probability of a driver accepting a gap is defined by logit model as $Y_i = \log\left(\frac{P_i}{1-P_i}\right) = \beta_0 + \beta_1 X_i$, Where β_o is a model constant and $\beta_1$is parameter of the specific independent variables. The predictor, X_i is the size of gap either accepted or rejected. The term $\log\left(\frac{P_i}{1-P_i}\right)$ in the above equation can be defined as the logit. Critical gap can be defined as the gap when the probability of acceptance of a gap size is 0.5.

Analysis

Data Summary

Table 1 below shows critical gaps, total gaps, and the proportion of gaps less than or equal to 12 seconds for each intersection by time period. As noted in the data reduction section, gaps greater than 12 seconds were not considered to reflect gap selection behavior and were not included in the analyses. As a result, only smaller gaps were utilized in the analyses. For instance, 327 gaps were

Table 1. Summary Statistics of Available Gap Data

Sites	Period	Treatment sites: Proportion of gaps < = 12 sec	Treatment sites: Accepted / rejected	Treatment sites: Total	Control sites: Proportion of gaps < = 12 sec	Control sites: Accepted / rejected	Control sites: Total
Chippewa	Before	0.18	5.29	327	0.49	1.39	311
Cottonwood		0.48	2.17	431	0.45	1.69	486
Isanti		0.56	1.39	456	0.37	2.43	381
McLeod		0.72	0.60	783	0.68	0.70	571
Pipestone		0.35	1.96	403	0.43	1.47	242
Chippewa	One month after	0.27	4.47	421	0.50	1.48	456
Cottonwood		0.45	2.28	305	0.42	1.87	342
Isanti		0.61	1.11	673	0.45	2.09	504
McLeod		0.64	0.71	747	0.70	0.59	608
Pipestone		0.45	1.20	368	0.46	1.11	114
Chippewa	12 months after	0.24	3.66	382	0.48	0.89	560
Cottonwood		0.46	1.86	438	0.46	1.62	430
Isanti		0.59	0.91	628	0.42	1.55	385
McLeod		0.28	0.65	536	0.70	0.56	757
Pipestone		0.04	27.50	228	0.24	3.04	202

Table 2. Critical Gap at Treatment Sites

Site	Method	Treatment: Before	Treatment: 1 month after	Treatment: 12 months after	Change: (1 month after) - (before)	Change: (12 months after) - (before)
Chippewa	Raffs'	9.80	7.70	11.00	−2.10	1.20
Chippewa	Logit	10.04	7.98	11.20	−2.06	1.16
Cottonwood	Raffs'	7.70	8.00	8.70	0.30	1.00
Cottonwood	Logit	7.74	8.53	8.89	0.80	1.16
Isanti	Raffs'	8.30	8.10	9.60	−0.20	1.30
Isanti	Logit	8.73	8.43	10.53	−0.31	1.79
McLeod	Raffs'	8.60	9.10	9.60	0.50	1.00
McLeod	Logit	9.58	9.84	10.60	0.26	1.02
Pipestone	Raffs'	10.00	11.12	10.50	1.12	0.50
Pipestone	Logit	9.35	10.46	10.21	1.11	0.86

reduced for the before period at the Chippewa treatment site and 18% were ≤ 12 seconds (59 gaps). Total indicates all gaps that were available and the proportion of smaller gaps is presented to show sample size utilized in the study. The table shows the ratio of total accepted to rejected gaps. As a result, Chippewa shows the proportion of accepted and rejected gaps as 5.29 indicating that the number of accepted gaps were five times more than the available rejected gaps. Similar information is provided for all time periods and intersections.

Critical Gap Analysis

As noted in the methodology section, critical gap was evaluated using both Raff's and Logit methods. Table 2 compares gap size for the different treatment sites across the three different time frames. As noted, the critical gap size decreased from 9.80 to 7.70 seconds (Raff's method) at the Chippewa treatment site one month after the installation of the ICWS while the critical gap increased to 11 seconds at 12 months after. Two sites showed a decrease in gap size at one month (–0.2 to 2.1 seconds) for both Raffs and the Logit model while two experienced an increase of 0.30–1.12 seconds. However, at the 12 months after period, all of the sites experienced an increase in gap size with both methods, and three of the four treatment sites had an increase in gap size of one or more seconds. This suggests that drivers adjusted to the treatment over time. It is assumed that an increase in critical gap size can be correlated to a positive safety benefit.

Table 3 shows the critical gaps at control sites during three different time frames. As indicated, control sites experienced changes similar to the treatment sites at one month after installation. Two control sites experienced a

Table 3. Critical Gap at Control Sites

Site	Method	Control			Change	
		Before	1 month after	12 months after	(1 month after) - (before)	(12 months after) - (before)
Chippewa	Raffs'	8.80	8.60	10.50	−0.20	1.70
Chippewa	Logit	9.70	9.60	11.40	−0.10	1.70
Cottonwood	Raffs'	8.60	9.55	9.40	0.95	0.80
Cottonwood	Logit	9.01	9.81	9.85	0.80	0.84
Isanti	Raffs'	8.80	8.10	10.00	−0.70	1.20
Isanti	Logit	8.86	8.28	10.05	−0.58	1.19
McLeod	Raffs'	8.60	9.10	8.90	0.50	0.30
McLeod	Logit	9.11	9.76	9.43	0.65	0.32
Pipestone	Raffs'	10.40	11.60	10.00	1.20	−0.40
Pipestone	Logit	12.04	13.35	11.45	1.31	−0.59

Note: As shown in Table 2, increases in a critical gap size of more than a second for most of the treatment sites shows the system itself has an effect on the treatment sites only after 12 months.

Table 4. Result from Paired Sample t-test

Sites	Method	Paired sample t-test			
		1 month after and before	Mean difference	12 months after and before	Mean difference
Treatment	Raff's	t(4) = −0.139, p = 0.896	−0.08	t(4) = 7.255, p = 0.002	1.00
	Logit	t(4) = −0.071, p = 0.947	−0.04	t(4) = 7.553, p = 0.002	1.19
Control	Raff's	t(4) = 0.988, p = 0.379	0.35	t(4) = 1.986, p = 0.118	0.72
	Logit	t(4) = 1.237, p = 0.284	0.42	t(4) = 1.767, p = 0.152	0.69

decrease in gap size of up to 0.7 seconds while two sites showed an increase of up to 1.3 seconds. At the 12 months after period, most sites experienced an increase of up to 1.7 seconds (results for both Raffs and Logit method). Pipestone showed a decrease in gap size of 0.40 and 0.59 seconds after installation.

However, no specific change or trend in gap size can be observed one month after. Similarly, Table 3 shows no specific trend or change in critical gap size either one month or 12 months after installation for control sites. These findings indicate no effect of the treatment sites on the control sites. A paired t-test was conducted separately at both treatment and control sites to test the significance of the effect of the system. Table 4 shows the results of a paired t-test (based on the p value and a level of significance of 5%) where the effect of the system at the treatment site was found to be significant 12 months after installation. No similar significance was found for the control sites.

Initially the intention of the control sites was to assess whether driver behavior changed overall and not just at the treatment intersections. Several studies had suggested that drivers were less likely to come to a full stop when ICWS were present since they did not perceive a need to stop. Although this does not necessarily present a safety hazard at the ICWS when the system is not activated, barring system malfunction, since no crossing vehicles were present. However, MnDOT was concerned that drivers who frequented intersections with ICWS would change their behavior overall and be less likely to make a full stop.

Since gap size did change at treatment but not control sites, it was assumed that no spillover effect resulted for gap selection and the control sites were then treated as regular control sites. In this manner, they can be used to assess behavior independent of the system.

In addition, separate paired sample t-tests were conducted between the critical gaps analyzed by the two different methods to check if the difference is significantly greater. Critical gaps during the before time period from Raff's method were compared with the Logit model method. Similary, critical gaps from one month and 12 months after installation were also compared with no significant difference found (p value near to one). Because of the general acceptance of logit model over the Raff method (*18*), only the Logit model method was used for further analysis: type of turn, type of vehicles, and type of stop.

Critical Gap by Type of Turn. The critical gap was compared by the type of turning movement made (right, through,

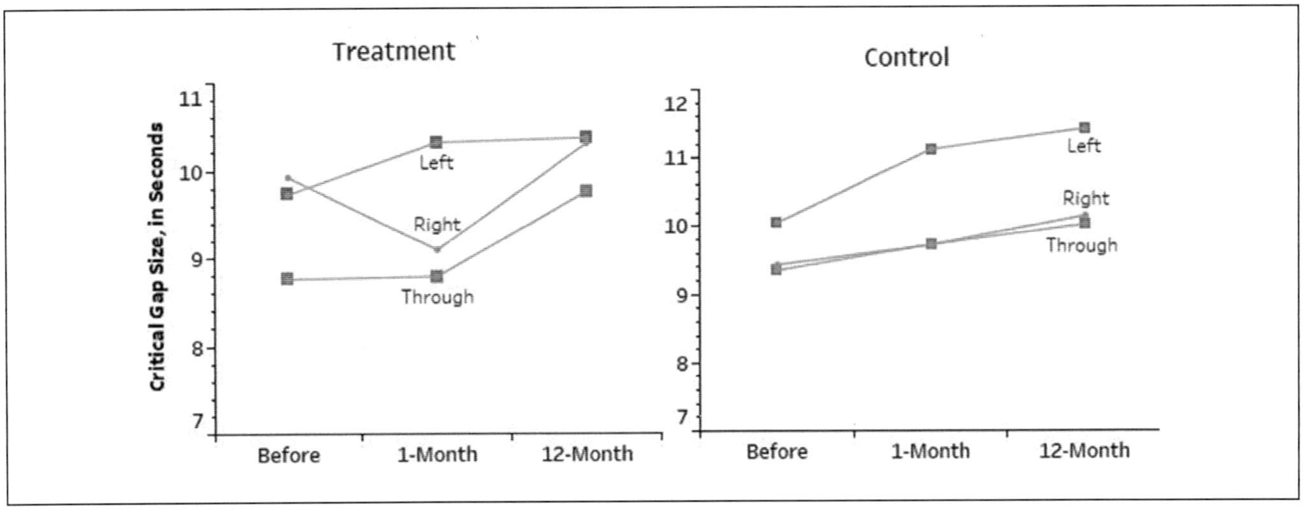

Figure 3. Overall critical gap by type of turn.

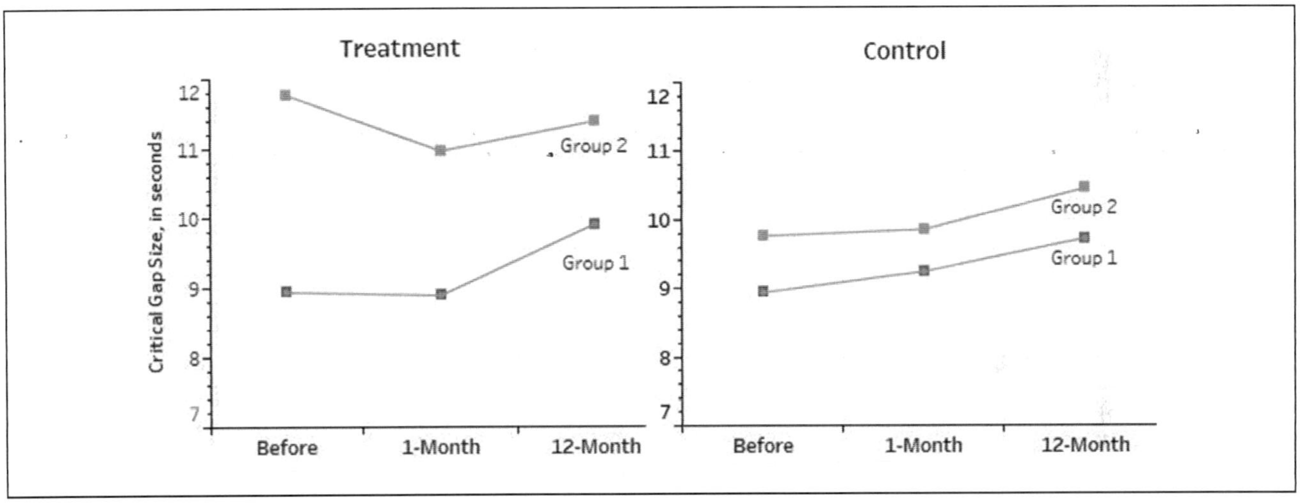

Figure 4. Overall critical gap by vehicle type.

and left). Due to sample size, data from all of the treatment sites were combined. Additionally, as mentioned above, only the logit model method was used for the analysis. Figure 3 shows the treatment site results for critical gap size by turning movement. For left turning vehicles, the critical gap increased by approximately 0.6 seconds at both one month and 12 months. For right turning vehicles, the critical gap size decreased by 0.8 seconds at one month after the treatment but increased by 0.4 seconds at 12 months. The critical gap for through vehicles did not change significantly at one month but increased by 1 second at 12 months. At control sites, the critical gap increased for all turning movements. The increase for left turns was more than 1 second and smaller increases were noted for through and right turns.

Critical Gap by Vehicle Type. Figure 4 shows the findings for critical gap by type of vehicle in Group 1, including passenger vehicles/motorcycles and Group 2, including larger vehicles such as trucks, buses, and farm vehicles. The critical gap for Group 1, at the treatment sites, showed no change between the before and one month after measurements and then increased by 1 second at the 12-month measurement. Group 2 (larger vehicles), at the treatment sites, had critical gaps that decreased by 1 second at one month after and 0.5 seconds at 12 months. As noted, passenger vehicles appeared to experience a much greater increase in critical gap size than truck drivers. This may suggest that they were either not relying on the system or were already taking their time in choosing gaps. The control site critical gaps, for both types of vehicle, increased by less than a second at both one month and 12 months after.

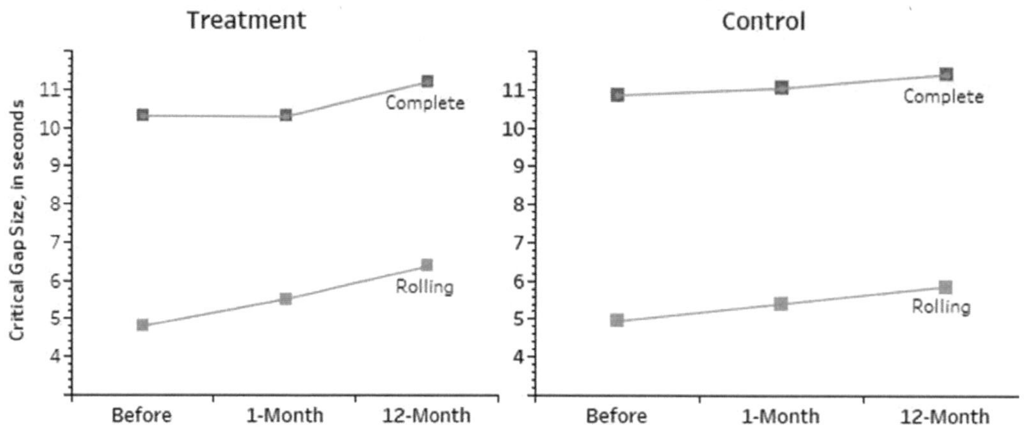

Figure 5. Overall critical gap by type of stop.

Critical Gap by Type of Stop. Figure 5 illustrates critical gap by type of stop (complete stop or rolling stop) with rolling stops associated with a much smaller critical gap than full stops. Some of this difference may be attributed to type of turning movement since vehicles who engage in a rolling stop may have been more likely to be turning right. For the treatment locations, the critical gap size for complete stops showed little change between the before and one-month measurements in contrast to an increase in gap size of 0.7 seconds for rolling stops at one month. At 12 months, the critical gap size increased for both types of stops (0.9 and 1.6 seconds for complete and rolling stops respectively). For the control sites, the increase in the critical gap size was very small comparatively. Critical gap size increased at 12 months but at a smaller rate than at treatment intersections.

Conclusion and Discussion

Analysis of critical gap size for the treatment sites shows no significant increases one month after installation and modest increases (0.5–1.79 second) 12 months after installation. The effectiveness of the system after 12 months indicates that drivers may take time to get familiar with the system. No significant change in critical gap was measured at the control sites.

Minor approach left turning vehicles responded to the ICWS system at one month and maintained an increased critical gap 12 months after installation. Through traffic did not show any response at one month but increased the gap size at 12 months. On the other hand, right turning vehicles experienced a decrease in critical gap at one month after installation but increased in gap at 12 months to similar levels as left turning vehicles. The decrease in gap size for right turning vehicles, one month after the installation of system, might be due to activation of the ICWS from vehicles approaching from opposing lanes to which right turning vehicles do not have to respond. For the control sites, the left turning vehicles increased critical gap sizes during both after periods and the right and through vehicles showed moderate increases of less than a second.

The analysis of critical gap size for the treatment sites by vehicle type shows little change for passenger vehicles at one month compared with an increase in gap by 12 months after installation. Larger vehicles (Group 2) at the treatment sites consistently had larger gap sizes, in contrast to Group 1. In this group, the gap did not change significantly over time. At control sites, both types of vehicle increased critical gap but by less than a second.

At the treatment sites, vehicles making a complete stop were found to be associated with larger critical gaps compared with a rolling stop. Vehicles with both types of stopping behavior were found to increase the size of the critical gap at 12 months after installation. However, one month after the installation of the system, vehicles who engaged in a rolling stop showed a critical gap of less than the system activation threshold of 6.5 seconds. This indicates that vehicles with non-stopping behavior had greater tendency to accept gaps even at the time of system activation. This might be due to the right-turning vehicles accepting gaps at the time the system activated from major street traffic approaching the intersection from opposite lanes, which has no effect on its movement. At control sites, increases in the critical gap were less than a second for both types of stop category.

Overall, the results show an effectiveness of the ICWS system in increasing critical gaps 12 months after installation. Both minor street left and through vehicles showed a consistent positive response to the system (increased gap size). Passenger vehicles had increased gap sizes at both one and 12 months. The effect of the ICWS at nearby control sites was not observed (no spillover effect).

Acknowledgments

The study would like to acknowledge the support of Minnesota Department of Transportation for funding this research project. The team would like to thanks Michael Kronzer, Cory Johnson, Brad Estochen, and members of the Technical Advisory Panel for their guidance and insight.

Author Contributions

The authors confirm contribution to the paper as follows: study conception and design: Hallmark, Hawkins, Knickerbocker, Thapa; data collection: Knickerbocker, Thapa; analysis and interpretation of results: Hallmark, Hawkins, Knickerbocker, Thapa; draft manuscript preparation: Hallmark, Hawkins, Knickerbocker, Thapa. All authors reviewed the results and approved the final version of the manuscript.

References

1. Intelligent Transportation System. Rural Unsignalized Intersections. Center for Transportation Studies. University of Minnesota. http://www.its.umn.edu/Research/FeaturedStudies/intersections/Accessed June 5, 2017.
2. Crowson, G., and J. Jackels. *Creek Consultants. Concept of Operations for Intersection Conflict Warning Systems (ICWS)*. ENTERPRISE Pooled Fund Study TPF-5(231). Michigan DOT (Administering State), Lansing, 2012.
3. Federal Highway Administration. *Intersection Collision Warning System. Tech Brief.* FHWA-RD-99-103. Federal Highway Administration, Washington, D.C., 1999.
4. Weidemann, R., T. Kwon, V. Lund, and B. Border. Determining the effectiveness of an advanced LED warning system for rural intersections. *Transportation Research Record: Journal of the Transportation Research Board,* 2011. 2250: 25–31.
5. Rakauskas, M., J. Creaser, M. Manser, J. Graving, and M. Donath. *Validation Study – On-Road Evaluation of the Stop Sign Assist Decision Support Sign.* CICAS-SSA Report # 5. University of Minnesota, 2009.
6. Ismail, H., T. M. Kwon, V. Lund, R. Ege, and A. Rindels. Development and Evaluation of an Advanced LED Warning System for Rural Intersections: Phase II (ALERT-2). 24th Annual Transportation Research Conference, St Paul, MN, 2013.
7. Simpson, C. L., and S. Troy. Safety Effectiveness of "Vehicle Entering When Flashing" Signs and Actuated Flashers at 74 Stop-Controlled Intersections in North Carolina. *Transportation Research Record: Journal of the Transportation Research Board,* 2013. 2384: 1–9.
8. Kwon, T. M., and H. Ismail. *Advanced LED Warning System for Rural Intersections: Phase 2 (ALERT-2)*. MN/RC 2014-10. Minnesota Department of Transportation, 2014.
9. Himes, S., F. Gross, K. Eccles, and B. Persaud. *Safety Evaluation of Intersection Conflict Warning Systems.* Publication No. FHWA-HRT-16-035. U.S. Department of Transportation. Federal Highway Administration, Washington, D.C., 2016.
10. Manser, M., and J. Creaser. Assessing driver behavioral adaptation to a rural intersection driver support system. Presented at 6th International Driving Symposium on Human Factors in Driver Assessment, Training and Vehicle Design, Lake Tahoe, Calif., 2011.
11. Madanat, S. M., M. J. Cassidy, and M. Wang. Probabilistic Delay Model at Stop-Controlled Intersection. *Journal of Transportation Engineering*, Vol. 120, No. I, 1994, pp. 21–36.
12. Kittelson, W. K., and M. A. Vandehey. Delay effects on driver gap acceptance characteristics at two-way stop-controlled intersections. *Transportation Research Record: Journal of the Transportation Research Board,* 1991. 1320: 154–159.
13. Miller, A. J. Nine estimators of gap-acceptance parameters. *Proc., 5th International Symposium on the Traffic Flow and Transportation*, June 1971, Berkeley, CA, pp. 215–235.
14. Patil, G. R., and S. Pawar. Temporal and Spatial Gap Acceptance for Minor Road at Uncontrolled Intersections in India. *Transportation Research Record: Journal of the Transportation Research Board,* 2014. 2461: 129–136.
15. Gattis, J. L., and S. T. Low. Gap acceptance at atypical stop-controlled intersections. *Journal of Transportation Engineering*, Vol. 125, 1999, pp. 201–207.
16. Goswamy, V., and G. H. Bham. Gap Acceptance Behavior in Mandatory Lane Changes under Congested and Uncongested Traffic on a Multi-lane Freeway. Presented at 86th Annual Meeting of the Transportation Research Board, Washington, D.C., 2007.
17. Raff, M. S., and J. W. Hart. *A Volume Warrant for Urban Stop Signs.* Eno Foundation for Highway Traffic Control, Saugatuck, Conn., 1950.
18. Fitzpatrick, K. Gaps accepted at stop-controlled intersections. *Transportation Research Record,* 1991, 1303: 103–122.

The Standing Committee on Traffic Control Devices (AHB50) peer-reviewed this paper (18-03508).

Article

TRR
JOURNAL OF THE TRANSPORTATION RESEARCH BOARD

Transportation Research Record
2018, Vol. 2672(21) 10–19

DOI: 10.1177/0361198118796709
journals.sagepub.com/home/trr
SAGE

Safety Effects of Turning Movement Restrictions at Stop-Controlled Intersections

Thanh Q. Le[1], Frank Gross[1], and Tim Harmon[1]

Abstract

This study evaluates the safety effectiveness of physical right-in-right-out (RIRO) operations compared with full turning movements at stop-controlled intersections. Geometric, traffic, and crash data from California were obtained for urban, three-legged, stop-controlled intersections with full movement and RIRO operations, as well as the downstream four-legged, stop-controlled or signalized intersections with full movement. A cross-sectional analysis provided estimates of the effects of turning movement restrictions while controlling for other differences between sites with RIRO and full movement. The aggregate results indicate reductions in total, all intersection-related, and fatal and injury intersection-related crashes at intersections with RIRO operations compared with full movement, with estimated crash modification factors of 0.55, 0.32, and 0.20, respectively. The reductions are statistically significant at the 95% confidence level for all crash types. Based on the disaggregate results, it does not appear that RIRO operations have different effects for different levels of traffic, design speed, or number of lanes. The analysis also examined the potential for crash migration from intersections where RIRO is implemented to the downstream intersection when determining the net benefits. The results indicate potential crash increases at downstream intersections, but many of the increases are not statistically significant at the 90% confidence level. Although the safety benefit-cost analysis suggests the strategy can be cost effective in reducing crashes at stop-controlled intersections, there is a need to analyze potential costs and benefits on a case-by-case basis with site-specific values.

Transportation agencies commonly use signs, pavement markings, or geometrics to prohibit or restrict turning movements with the objective of improving safety or operations. Unprotected left-turn movements crossing a conflicting direction of traffic present a risk for crashes. Right turns at most stop-controlled intersections are essentially merging movements and do not present the same level of crash risks as left turns. Restricted or prohibited turning movements reduce the number of turning conflict points at intersections, which is generally known to reduce crash risk (*1*).

The turning operations at most stop-controlled intersections can usually be categorized into one of the following three groups:

- Full movement.
- Left turn from major road only.
- Right-in-right-out (RIRO).

Full movement implies no turning restrictions; most stop-controlled intersections operate with full movement. Left turn from major road only prohibits left turns out of the minor road while allowing left turns from the major road, such as a restricted crossing U-turn intersection. RIRO eliminates left turns into and out of the minor road. A positive or curbed median barrier on the major road is the most common strategy for implementing RIRO at stop-controlled intersections and driveways. Medians physically block left turns into and out of the intersecting street. Figure 1 presents the image of a stop-controlled intersection with RIRO operations.

As with all access management techniques, agencies must strike a balance between the safety and operational efficiency of intersections while maintaining access to properties along, and adjacent to, the roadway. Although restricting turns is expected to provide a safety improvement in most cases, there is limited information available about the quantitative safety effectiveness of these practices and the effects on downstream

[1]Vanasse Hangen Brustlin, Raleigh, NC

Corresponding Author:
Address correspondence to Thanh Q. Le: tle@vhb.com

Figure 1. RIRO stop-controlled intersection with positive median barrier.

intersections. This study serves to address the need for research into the safety effectiveness of turning movement restrictions at stop-controlled intersections. This research was conducted under the Federal Highway Administration's (FHWA) Development of Crash Modification Factors program for the Evaluation of Low Cost Safety Improvements Pooled Study.

Literature Review

The literature review focused on the safety effects of RIRO operations, which are most commonly implemented with a raised median preventing all left turns. Most or all evaluations to date have examined corridor and segment impacts of installing raised medians rather than the effects of turning restrictions at the treated stop-controlled intersections and downstream intersections. The following provides a summary of the salient research related to specific strategies.

Schultz et al. indicated that the presence of a raised median corresponded to a reduction of 1.23 crashes per million vehicle miles traveled (*2*). In addition, raised medians were negatively correlated with right-angle collisions. Gluck et al., in NCHRP Report 420, investigated the relationship between median type and crash rates (*3*). NCHRP Report 395 compares different outcomes from several crash prediction models developed by different researchers regarding midblock left turns (*4*). A composite finding suggested that, in general, a raised median is safer than undivided roadways, especially on roads with above 20,000 vehicles per day. Eisele and Frawley investigated the relationship between access density and crash rate for raised median and non-raised median corridors separately (*5*). Both relationships were positively correlated, but the trend was slightly steeper for non-raised median corridors than raised median corridors. The researchers concluded that the reduced slope of the regression line for raised median corridors demonstrates there are relatively lower crash rates in corridors with raised medians as a result of the reduced conflict points. A study with data from seven States by Hallmark et al. suggested raised medians reduced crashes by at least 40 percent in urban settings (*6*).

Research Objective

This research examined the safety impacts of physically restricting turning movements to implement RIRO operations at stop-controlled intersections in California. The objective was to estimate the safety effectiveness of this strategy as measured by crash frequency. Target crash types included the following:

- Total: all crashes within 100 ft of intersection (all types and severities combined).
- Intersection-related: all crashes within 100 ft of intersection defined as "intersection-related" by the reporting officer (all types and severities combined).
- Fatal and injury: all injury crashes within 100 ft of intersection defined as "intersection-related" by the reporting officer (K, A, B, and C injuries on KABCO scale).
- Multi-vehicle: all multiple-vehicle crashes within 100 ft of intersection defined as "intersection-related" by the reporting officer (all types and

severities combined). Intersection-related crashes within 100 ft of the intersections all included multiple vehicles, so the project team dropped the multi-vehicle category from the remainder of the analysis, as it was redundant.

A further objective was to address the following questions:

- Do effects vary by lane configuration of intersection (i.e., four major road lanes and two minor road lanes versus six major road lanes and two minor road lanes)?
- Do effects vary by level of traffic volume?
- Do effects vary by design speed on the major road route?
- Do effects vary by the type of traffic control at downstream intersections (i.e., signalized or stop-controlled)?
- Do effects vary by the presence of turn lanes at downstream intersections?

The evaluation of overall effectiveness included the consideration of the installation costs and crash savings in terms of a benefit-cost ratio.

Methodology

The evaluation used a cross-sectional study design. At the most basic level, the safety effect is estimated by taking the ratio of the average crash frequency for two groups, one with the treatment and the other without the treatment. In cross-sectional studies, the two groups of sites should be similar in all regards except for the presence of the treatment. This segregation of sites is difficult to accomplish in practice. Propensity score matching was used to identify similar sites with and without treatment, and multivariable regression modeling was used to control for other characteristics that vary among sites. Detailed discussions of propensity score matching and its application in traffic safety research are available in papers by Rosenbaum and Rubin and Sasidharan and Donnell (*7*, *8*).

The project team developed multivariable, negative binomial regression models to explain the statistical relationships between the dependent variable (i.e., crash frequency) and a set of predictor variables. The research team considered many predictor variables including treatment presence, traffic volume, and other roadway characteristics. The team estimated regression coefficients during the modeling process for each predictor variable. The coefficients represent the expected change in crash frequency as a result of a unit change in the predictor variable with all else being equal. One concern was the possibility of site-selection bias if agencies installed turning movement restrictions to address safety issues. Propensity score matching helps to overcome potential site-selection bias.

Data Collection

The majority of data for this study were collected under a previous project funded by the FHWA entitled *Safety Evaluation of Access Management Policies and Techniques* (*9*). The current study relied on geographic information system (GIS) files compiled under the previous effort to identify candidate intersections for this evaluation. The GIS files provided the location and type of turning restriction (i.e., full movement or RIRO) for intersections across California.

The research team supplemented the existing GIS files with additional data from the Highway Safety Information System (HSIS) database. The HSIS roadway inventory provided number of lanes, lane width, shoulder width, design speed, average annual daily traffic (AADT), and other geometric characteristics on the major road roads. The intersection inventory from HSIS includes routes, county numbers, and milepost on the major road of all intersections. The HSIS inventory also provided AADT on the minor road. HSIS data were verified with aerial imagery using Google Earth.

Milepost, county, and route number were used to identify and link crashes from the HSIS crash data files to each intersection. All crashes that occurred within a 200-ft influence zone from the center of the intersection (i.e., 100 ft upstream and 100 ft downstream) were included in the analysis. The research team used crash report codes to determine which were reported as "intersection-related" crashes.

In addition to collecting data for stop-controlled intersections with and without RIRO, data from downstream intersections were collected to model possible crash migration from an intersection where RIRO was implemented to the nearest location where vehicles make U-turns. In all cases, the downstream intersection was designated as the closest downstream intersection following the RIRO or full movement intersection where U-turns were allowed (i.e., if the immediate downstream intersection had U-turn restrictions, the next following intersection was selected). Only the closest upstream intersection from each downstream intersection was included on either side of any road.

The research team collected data for 333 candidate stop-controlled intersections and 202 downstream intersections along corridors of various facility types. During preliminary data analysis, the team decided to retain only locations with four or six lanes on the major road. There were too few sites within the available data sets

Table 1. Data Summary for Urban, Three-Legged, Stop-Controlled Intersections with RIRO Operations

Variable	Sites	Mean	Std. dev.	Min	Max
Major road AADT	58	38,724	11,997	13,433	75,000
Minor road AADT	58	519	510	51	2,600
Major road 6-lane indicator (1 if 6 lanes, 0 if 4 lanes)	58	0.586	0.497	0	1
50 + mph indicator (1 if 50 + mph, 0 otherwise)	58	0.224	0.421	0	1
Total crashes	58	2.586	2.555	0	9
Intersection-related crashes	58	0.638	0.968	0	4
Fatal and injury, intersection-related crashes	58	0.190	0.438	0	2

Table 2. Data Summary for Urban, Three-Legged, Stop-Controlled Intersections with Full Movement

Variable	Sites	Mean	Std. dev.	Min	Max
Major road AADT	80	34,271	11,719	9,940	75,000
Minor road AADT	80	765	759	51	3,650
Major road 6-lane indicator (1 if 6 lanes, 0 if 4 lanes)	80	0.288	0.455	0	1
50 + mph indicator (1 if 50 + mph, 0 otherwise)	80	0.500	0.503	0	1
Total crashes	80	4.163	3.777	0	17
Intersection-related crashes	80	2.025	2.658	0	11
Fatal and injury, intersection-related crashes	80	1.125	1.618	0	8

having physical RIRO along corridors with two lanes, five lanes, or more than six lanes to draw meaningful conclusions on those facility types. Further, based on the literature review, prior studies indicate very little, if any, benefit of implementing RIRO at intersections and driveways where left-turning traffic only crosses one lane. All intersecting segments at the three-legged RIRO or full movement sites serve two-way traffic. The distance from the stop-controlled intersection of interest to the downstream intersection, the presence and length of left-turn right-turn lanes at the downstream intersection were also identified and measured. The research team did not explicitly identify whether downstream intersection minor roads served one-way or two-way traffic; regardless, each downstream intersection allowed the necessary U-turn movement. The signal phasing for downstream intersections is unknown and may include permissive, permissive-protected, and protected left-turn phasing.

The final dataset included 138 stop-controlled intersections with a mix of physical RIRO and full movement along four lane and six lane corridors. The downstream intersection dataset included 109 intersections with a mix of stop- and signal-control. The number of downstream intersections is smaller because some three-legged, stop-controlled intersections have the same downstream intersection. This case occurs when two upstream intersections (i.e., three-legged, stop-controlled) are located on opposite sides and opposite approaches to a common four-legged intersection. When two stop-controlled intersections share a downstream intersection, the project team confirmed the type of restriction (i.e., RIRO or full movement) is the same for the two parent intersections.

The project team collected and aggregated three years of data for the analysis. Tables 1 and 2 present a data summary for the 138 urban, three-legged, stop-controlled intersections included in the primary analysis. Table 1 presents a summary of 58 locations with physical RIRO and Table 2 presents a summary of 80 locations with full movement.

Tables 3 and 4 present data summaries for 109 intersections downstream from the primary study intersections. Table 3 presents a summary of 48 intersections downstream from intersections with RIRO operations and Table 4 presents a summary of 61 intersections downstream from intersections with full movement. Indicator variables are either 0 or 1, representing the absence or presence of the characteristic, respectively. The mean value of an indicator variable indicates the proportion of sites with the attribute present (indicator value of 1). For example, the six-lane indicator in Table 1 has a mean value of 0.586. This implies that 58.6% of locations have six lanes on the major road (indicator value = 1) and 41.4% of locations have four lanes (indicator value = 0). Similarly, the mean value of the signalized indicator in Table 3 is 0.771, indicating that 77.1% of the sample is signalized intersections.

Safety Performance Function Development

The safety performance function (SPF) has the following general functional form:

$$N = ML\_AADT^{\beta_1} \times XST\_AADT^{\beta_2} \times e^{(\beta_0 + \beta_i \times X_i)}$$

Table 3. Data Summary for Downstream Intersections of Locations with RIRO Operations

Variable	Sites	Mean	Std. dev.	Min	Max
Major road AADT	48	39,148	12,723	22,010	75,000
Minor road AADT	48	6,686	9,656	201	56,000
Signalized indicator (1 if signalized, 0 otherwise)	48	0.771	0.425	0	1
Major road 6-lane indicator (1 if 6 lanes, 0 if 4 lanes)	48	0.563	0.501	0	1
Total crashes	48	18.146	24.163	0	107
Intersection-related crashes	48	6.563	9.079	0	50
Fatal and injury, intersection-related crashes	48	2.771	3.508	0	12

Table 4. Data Summary for Downstream Intersections of Locations with Full Movement

Variable	Sites	Mean	Std. dev.	Min	Max
Major road AADT	61	34,573	12,489	8,867	75,000
Minor road AADT	61	3,918	5,459	51	25,390
Signalized indicator (1 if signalized, 0 otherwise)	61	0.557	0.501	0	1
Major road 6-lane indicator (1 if 6 lanes, 0 if 4 lanes)	61	0.295	0.460	0	1
Total crashes	61	8.475	7.997	0	36
Intersection-related crashes	61	3.213	3.843	0	21
Fatal and injury, intersection-related crashes	61	1.705	1.986	0	8

Table 5. Results for Urban, Three-Legged, Stop-Controlled Intersections with RIRO Compared with Full Movement

Variable	Total	Intersection- related	Fatal & injury
Observed crashes per site-year with RIRO	0.86	0.21	0.06
Observed crashes per site-year with full movement	1.39	0.68	0.38
Estimate of CMF	0.55[a]	0.32[a]	0.20[a]
Standard error of CMF	0.09	0.08	0.07

[a]Indicates statistically significant results at the 95% confidence level.

The variable definitions included in the final crash prediction models are as follows:

- N = predicted number of crashes.
- ML_AADT = AADT on the major road (two-way, vehicles/day). Traffic volume estimates are from HSIS, representing the bidirectional AADT for the roadway section.
- XST_AADT = AADT on the minor road (two-way, vehicles/day). Traffic volume estimates are from HSIS, representing the bidirectional AADT for the roadway section.
- β_0, β_1, β_2, β_i = model parameters to be estimated.
- X_i = explanatory variables representing the geometric and operational characteristics of the intersections.

The project team developed crash prediction models separately for urban, three-legged, stop-controlled intersections and downstream four-legged stop-controlled and signalized intersections. In the statistical model development process, the team tested potential variables representing the distance between the RIRO and downstream intersections, the presence of turning lanes at the downstream intersections. However, these variables did not contribute to better statistical fit for the model, so in the end they were left out of the final models. Readers are recommended to refer to the full report for the detailed discussion of the SPFs and model parameters (*10*).

Analysis Results and Discussion

Table 5 presents the estimated CMFs and related standard errors for each target crash type at the stop-controlled intersections with RIRO compared with similar intersections with full movement. The aggregate results indicate reductions for all crash types analyzed: total, all intersection-related, and fatal and injury intersection-related crashes. The reductions are statistically significant at the 95% confidence level for all crash

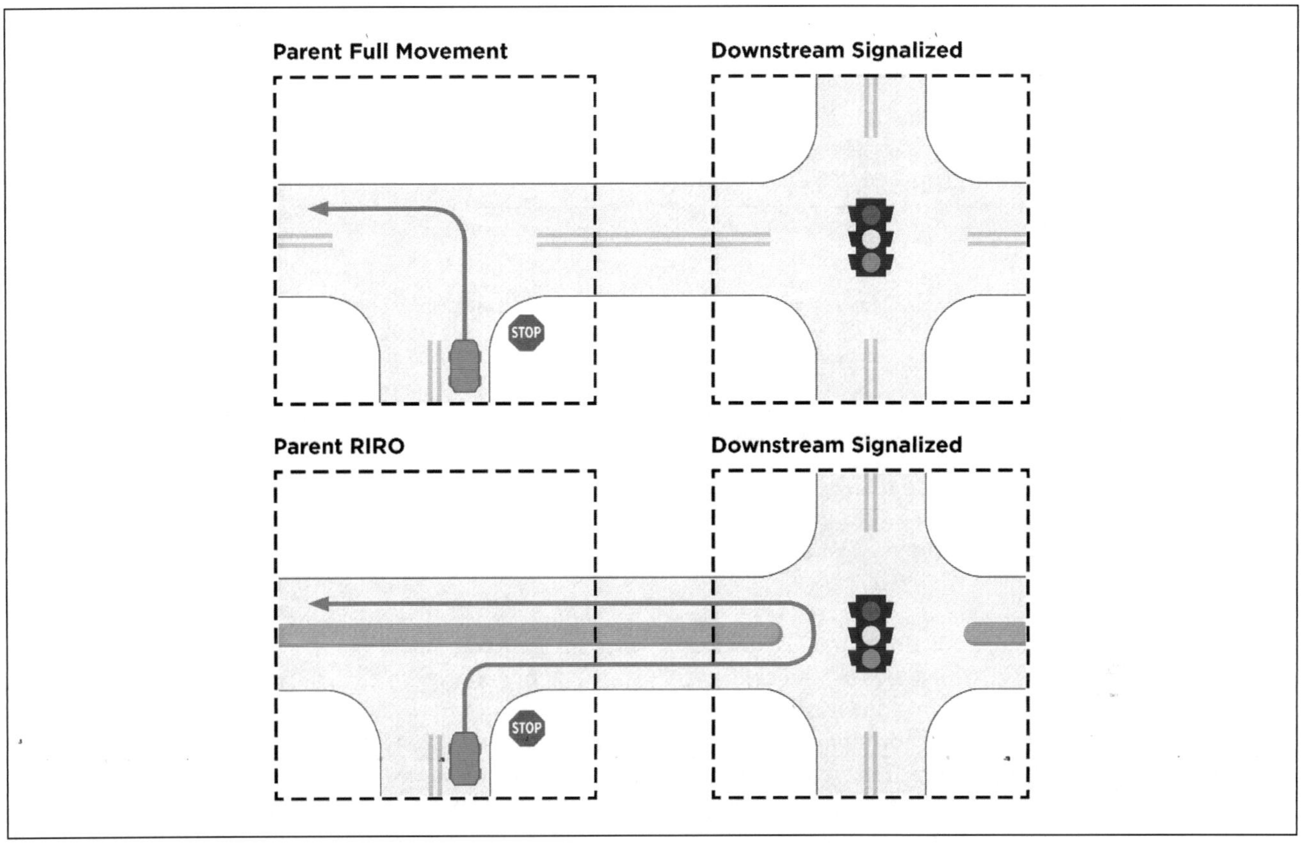

Figure 2. Illustration of upstream and downstream full movement intersections compared with upstream RIRO and downstream full movement intersections.

types. The CMFs for total, all intersection-related, and fatal and injury intersection-related crashes are 0.55, 0.32, and 0.20, respectively. It is important to note that these CMFs were not computed by simple division of observed crashes for sites with full movement and sites with RIRO. The CMFs and standard errors presented in this table were estimated from the parameters of crash prediction models.

Crash migration is a potential issue related to the physical restriction of turning movements at a given access point. This occurs when the turning traffic is shifted from the treated location to the downstream location to facilitate a U-turn to complete the previous movement, and likewise crashes at the treated site are shifted to another site. Although RIRO turning restrictions eliminate left-turn crashes at the treatment location, there is potential to increase crashes from more U-turning traffic at the next downstream intersections that allow U-turns.

Figure 2 illustrates the relocation of direct left turns at a full movement stop-controlled intersection to a downstream intersection when the target intersection is converted to RIRO. To estimate the change in safety performance, there is a need to understand the combined safety performance of the full movement stop-controlled intersection and downstream intersection compared with the safety performance of the RIRO stop-controlled intersection along with the downstream intersection.

In this analysis, the team used an interaction term between the RIRO indicator for the upstream intersection and the signal control indicator for the downstream intersection. This revealed differences in crash migration effects by traffic control type (i.e., signalized vs. stop-controlled) at the downstream U-turn location. Tables 6 and 7 present the estimated CMFs and related standard errors for each target crash type and traffic control type combination at the downstream intersections. The CMFs represent the difference in crashes at intersections downstream from RIRO intersections compared with intersections downstream from full movement intersections.

Table 6 presents the results for downstream intersections with signal control. The reductions are not statistically significant at the 90% confidence level; however, the estimates show potential for increases in total and intersection-related crashes at downstream signalized intersections. The CMFs for total, all intersection-related, and fatal and injury intersection-related crashes are 1.10, 1.02, and 0.94, respectively.

Table 6. Results for Urban Signalized Intersections Downstream from Stop-Controlled Intersections with RIRO Compared with Full Movement

Variable	Total	Intersection-related	Fatal & injury
Estimate of CMF (parent RIRO=1 and downstream SIGNAL=1)	1.10	1.02	0.94
Standard error of CMF (parent RIRO=1 and downstream SIGNAL=1)	0.20	0.24	0.26

Table 7. Results for Urban Stop-Controlled Intersections Downstream from Stop-Controlled Intersections with RIRO Compared with Full Movement

Variable	Total	Intersection-related	Fatal & injury
Estimate of CMF (parent RIRO=1 and downstream SIGNAL=0)	1.64[b]	2.55[a]	1.56
Standard error of CMF (parent RIRO=1 and downstream SIGNAL=0)	0.33	0.39	0.45

[a]Indicates statistically significant result at the 95% confidence level.
[b]Indicates statistically significant result at the 90% confidence level.

Table 7 presents the results for downstream intersections with stop control. All three CMFs indicate the potential for increased crashes at downstream stop-controlled intersections. The CMFs for total, all intersection-related, and fatal and injury intersection-related crashes are 1.64, 2.55, and 1.56, respectively. Among these, the CMF estimate for total crash is statistically significant at 90% and intersection-related crashes are statistically significant at 95% confidence levels.

A conservative analysis would include the potential increases at downstream intersections (i.e., even those not statistically significant) in the estimation of the net benefit of RIRO implementation.

Comparing the CMFs, the results indicate smaller potential changes for all crash types at downstream signalized intersections relative to downstream stop-controlled intersections. Further, the CMFs for downstream signalized intersections are not statistically significant at the 90% confidence level, and two of the three CMFs for downstream stop-controlled intersections are statistically significant at the 90% confidence level.

The research team considered several variables in a disaggregate analysis, including major road and minor road traffic volume, number of major road lanes, and design speed. Based on the disaggregate results, it does not appear that RIRO operations have different effects for different levels of traffic on both major road and minor road, design speed, or number of lanes on the major road.

Safety Benefit-Cost Analysis

The research team conducted a safety benefit-cost (B/C) analysis of a hypothetical scenario to estimate the cost-effectiveness, measured with a benefit-cost ratio, of installing curbed median to implement RIRO operations at a previously full movement stop-controlled intersection.

For estimating project costs in this analysis, the research team assumed the project would consist of constructing a 4-ft wide median at an average cost of \$6 per square ft (*11*). Given these assumptions, the construction cost is approximately \$24 per linear foot or \$126,720 per mile. The analysis assumes no roadway widening or other reconstruction to facilitate the installation of the median. A 1,210 ft median was analyzed, which represents the average distance between the centers of signalized intersections along corridors in the study data set minus 100 ft to account for the paved intersection areas. Thus, the average cost per installation between contiguous signalized intersections is approximately \$26,500. Annual average costs for routine mowing, sweeping, and minor curb and gutter repairs were assumed to be negligible along the 1,210-ft section.

Safety benefits were determined by multiplying the total crash CMFs (i.e., 0.20 for converting full movement to RIRO and 0.94 for the effect at the downstream signalized intersection) by the average crash frequency at each intersection (i.e., 3.0 fatal and injury crashes per site per year at target full movement three-legged, stop-controlled intersections and 5.0 fatal and injury crashes per site per year at downstream signalized intersections). The net difference between the estimated crash frequencies with and without RIRO operations of 2.7 fatal and injury crashes per year was multiplied by an average fatal and injury crash cost of \$391,280 in 2015 dollars (*12*, *13*). A service life of 20 years represents the period until significant maintenance or reconstruction of the median is needed.

The B/C ratio is determined by the total value of safety benefits divided by the project costs. A sensitivity

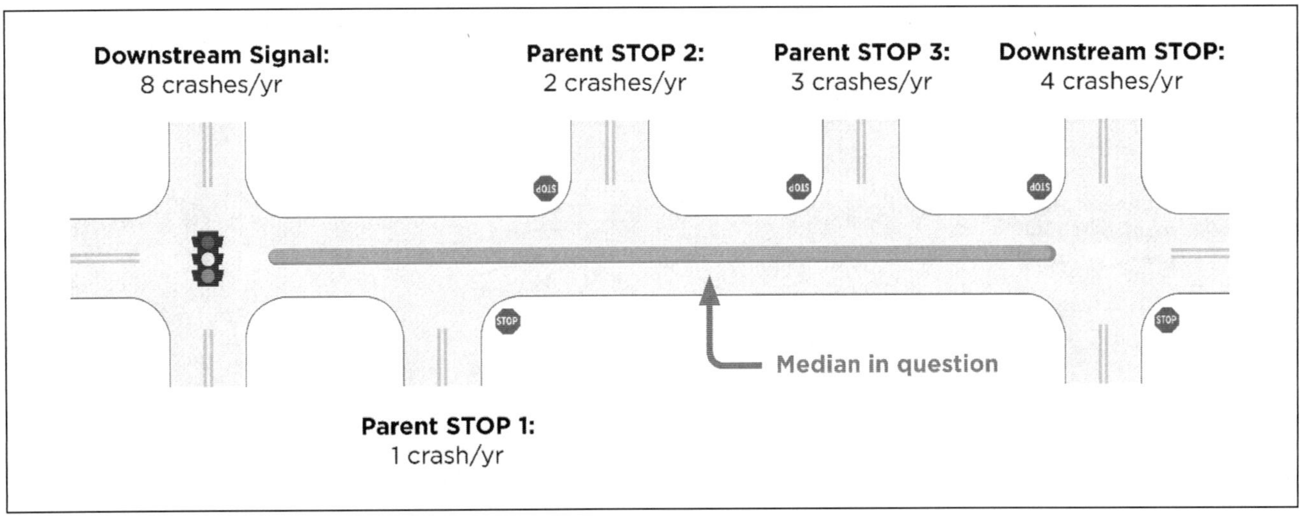

Figure 3. Illustration of intersections converted from full movement to RIRO.

analysis was conducted to determine the potential bounds of B/C ratio under various values of a statistical life, with lower and upper bounds at 0.55 and 1.38 times the 2015 value, respectively (*13*).

The safety B/C analysis suggests that implementing RIRO operations with a curbed median—with reasonable assumptions on cost, service life, and the value of a statistical life—can be cost effective for reducing crashes, with the B/C ratio potentially ranging from 5.4:1 to 13.5:1. The research team recommends conducting a new safety B/C analysis to determine if improving RIRO operations would be cost effective using the actual crash frequency of the site, proposed barrier type and other construction costs, and with site-specific corridor configurations where proposed projects are considered.

Example Corridor Application

The research team also considered the application of the CMFs along a hypothetical corridor with several previously full movement stop-controlled intersections. Figure 3 illustrates the example section.

This analysis uses the total crash CMFs of 0.55, 1.10, and 1.64 from Table 5 through Table 7 for converting full movement stop-controlled intersections to RIRO, the effect at downstream signalized intersections, and the effect at downstream stop-controlled intersections, respectively. To estimate the safety effect at a downstream intersection, an analyst must apply the applicable CMF once for each upstream intersection.

Table 8 shows the average total crash frequency per year prior to treatment, the applicable CMFs, estimated crashes per year after treatment (i.e., CMF times estimated crashes per year before treatment), and estimated change in crashes per year (i.e., estimated crashes per year after minus average crashes per year before treatment) at each intersection.

Although the target crashes at intersections with RIRO implementation are reduced, the net effect of installing a median to implement RIRO in this example is an increase of 1.54 total crashes per year across all sites. Therefore, the example treatment is not justified because of the estimated increase in crashes.

Summary and Conclusion

The objective of this study was to evaluate the safety effectiveness, as measured by crash frequency, of physical turning movement restrictions at urban, three-legged, stop-controlled intersections. The study compared RIRO with full movement access using data from California to examine the effects on total, intersection-related, and fatal and injury intersection-related crashes. The aggregate results indicate reductions for all crash types analyzed at urban, three-legged, stop-controlled intersections with RIRO compared with full movement. All reductions are statistically significant at the 95% confidence level.

The analysis of downstream intersections indicates potential crash migration leads to an increase in crashes at immediate downstream U-turn locations from RIRO intersections. There are differences in crash migration effects depending on the downstream traffic control type. The CMFs for downstream locations with signal control are close to 1.0 and the estimates are not statistically significant at the 90% confidence level. There is relatively low chance of increased crashes for downstream signalized intersections, although this may not be the case for all sites given high traffic volumes or other site conditions. Conversely, the results indicate likely increases at downstream stop-controlled intersections for all crash types analyzed. The increases are statistically significant

Table 8. Applying CMFs to a Corridor Scenario

Intersection	Average crashes per year before treatment	CMF	Estimated crashes per year after treatment	Estimated change in crashes per year
STOP 1 – Upstream from four-legged STOP	1.0	0.55	0.55	−0.45
STOP 2 – Upstream from four-legged signal	2.0	0.55	1.10	−0.90
STOP 3 – Upstream from four-legged signal	3.0	0.55	1.65	−1.35
Downstream four-legged STOP	4.0	1.64	6.56	+ 2.56
Downstream four-legged signal	8.0	1.21[a]	9.68	+ 1.68
Total	18.0	N/A	19.54	+ 1.54

[a]CMF multiplied twice (i.e., 1.10 × 1.10 = 1.21) to reflect two upstream intersections.

at 90% and 95% confidence levels for total and total intersection-related crashes, respectively. The disaggregate analysis indicated no statistically significant differences in effects for major road traffic volume, minor road traffic volume, and design speed.

The safety B/C analysis results suggest the strategy can be cost effective in reducing crashes at stop-controlled intersections. However, there is a need to analyze potential costs and benefits on a case-by-case basis with site-specific values, owing to potential crash migration causing projects to be unjustified.

Future research needs include the opportunity to evaluate similar data or validate the results with data from other States, the exploration of safety effects related to pedestrian crashes and other collision types, and expanding the analysis to include other facility types such as two-lane roads or larger facilities. Actual operating speed data would help improve the crash prediction models. There is also a need to consider the actual shift in turning movements from the parent RIRO location to the downstream U-turn locations.

Acknowledgments

This research was funded by the Federal Highway Administration, Office of Safety Research and Development under the Development of Crash Modification Factors program for the Evaluation of Low Cost Safety Improvements Pooled Fund Study. The FHWA Program and Task Manager for the project was Roya Amjadi.

Author Contributions

The authors confirm contribution to the paper as follows: study conception and design: Frank Gross, Thanh Q. Le; data collection: Thanh Q. Le; analysis and interpretation of results: Thanh Q. Le, Frank Gross, Tim Harmon; draft manuscript preparation: Thanh Q. Le. All authors reviewed the results and approved the final version of the manuscript.

References

1. Simodynes, T., T. Welch, and M. Kuntemeyer. *Effects of Reducing Conflict Points on Reducing Accidents* (Abstract Only). Third National Access Management Conference, Federal Highway Administration, Washington, D.C., 2000, p. 141.
2. Schultz, G. G., K. T. Braley, and T. Boschert. Correlating Access Management with Crash Rate, Severity, and Collision Type. Presented at 87th Annual Meeting of the Transportation Research Board, Washington, D.C., 2008.
3. Gluck, J. S., H. S. Levinson, and V. Stover. *NCHRP Report 420: Impact of Access Management Techniques.* Transportation Research Board, National Research Council, Washington D.C., 1999.
4. Bonneson, J. A., and P. T. McCoy. *NCHRP Report 395: Capacity and Operational Effects of Midblock Left-Turn Lanes.* Transportation Research Board of the National Academies, Washington, D.C., 1997.
5. Eisele, W. L., and W. E. Frawley. Estimating the Safety and Operational Impact of Raised Medians and Driveway Density: Experiences from Texas and Oklahoma Case Studies. *Transportation Research Record: Journal of the Transportation Research Board*, 2005. 1931: 108–116.
6. Hallmark, S., D. Plazak, E. Fitzsimmons, K. Hoth, and H. Isebrands. *Toolbox to Assess Tradeoffs between Safety, Operations, and Air Quality for Intersection and Access Management Strategies: Final Report.* Iowa State University, Center for Transportation Research and Education; Federal Highway Administration; Midwest Transportation Consortium, Ames, IA, 2008, pp. 1–121.
7. Rosenbaum, P., and D. Rubin. *The Central Role of the Propensity Score in Observational Studies for Causal Effects.* Biometrika, Las Vegas, NV, Vol. 70, No.1, 1983, pp. 41–55.
8. Sasidharan, L., and E. Donnell. Application of Propensity Scores and Potential Outcomes to Estimate Effectiveness of Traffic Safety Countermeasures: Exploratory Analysis Using Intersection Lighting Data. *Accident Analysis and Prevention*, Vol. 50, 2013, pp. 539–553.
9. Gross, F., C. Lyon, B. Persaud, J. Gluck, M. Lorenz, and S. Himes, *Safety Evaluation of Access Management Policies and Techniques.* FHWA-HRT-15-038. Federal Highway Administration, Washington, D.C., 2015.
10. Le, T., F. Gross, T. Harmon, K. Eccles, and B. Persaud. *Safety Evaluation of Turning Movement Restrictions at Stop-Controlled Intersections.* FHWA-HRT-17-064. Federal Highway Administration, Washington, D.C., 2018.

11. Bushell, Max, A., B. W. Poole, C. V. Zegeer, and D. A. Rodriguez. *Costs for Pedestrian and Bicyclist Infrastructure Improvements.* UNC Highway Safety Research Center, Chapel Hill, NC, 2013.
12. Council, F., E. Zaloshnja, T Miller, and B. Persaud. *Crash Cost Estimates by Maximum Police-Reported Injury Severity within Selected Crash Geometries.* FHWA-HRT05-051. Federal Highway Administration, Richmond, VA, 2005.
13. U.S. Department of Transportation. *Guidance on Treatment of the Economic Value of a Statistical Life (VSL) in U.S. Department of Transportation Analyses—2015 Adjustment.* Memorandum, Washington, D.C., June 17, 2015. https://cms.dot.gov/sites/dot.gov/files/docs/VSL2015_0.pdf. Accessed May 9, 2016.

The Standing Committee on Traffic Control Devices (AHB50) peer-reviewed this paper (18-03753).

Article

TRR
JOURNAL OF THE TRANSPORTATION RESEARCH BOARD

Transportation Research Record
2018, Vol. 2672(21) 20–29

DOI: 10.1177/0361198118794289
journals.sagepub.com/home/trr
SAGE

Safety Effects of Flashing Yellow Arrows Used in Protected Permitted Phasing: Comparison of Full Bayes and Empirical Bayes Results

Justice Appiah[1], F. Adam King[1], Michael D. Fontaine[1], and Benjamin H. Cottrell, Jr.[1]

Abstract

Using the flashing yellow arrow (FYA) signal indication for the permissive portion of protected-permissive left-turn (PPLT) phasing has become an increasingly popular treatment for left-turn signals as drivers are believed to understand the FYA better than the traditional circular green indication. A before-and-after safety evaluation of deploying FYA at PPLT signals at 28 intersections in Virginia was conducted. Each of the study intersections had FYA for the permitted portion of the phase on at least one left-turn approach. The focus was on left-turns that operated in the protected-permissive mode (with circular green indication for the permissive portion) before being converted to PPLT operations with the FYA indication for the permissive portion (PPLT-FYA). Crash records from before and after the activation of FYA were compared using both the full Bayes and empirical Bayes approaches. The results indicate that using the FYA signal indication instead of the circular green indication had a statistically significant effect in reducing overall frequency and severity of crashes. For the intersections studied in this research, total crashes reduced by 12% following conversion from PPLT to PPLT–FYA. The results also indicated that the full Bayes approach to safety effectiveness evaluation can, at a minimum, provide similar results to the well-established empirical Bayes approach. The 95% credible intervals for the expected crash reduction rates estimated with the full Bayes method were generally narrow, suggesting a good degree of confidence in the estimates.

Permissive left-turn phases at signalized intersections must be clearly communicated to drivers to ensure that drivers understand who has the right-of-way. If drivers misjudge whether they have the right-of-way when making a turn, they risk collision with vehicles traveling in the opposite direction. In a 2003 *National Cooperative Highway Research Program* (NCHRP) report, Brehmer et al. recommended that the flashing yellow arrow (FYA) signal display be used to indicate permissive left-turns because it was better understood by drivers than the traditional circular green (*1*). The FYA was subsequently included in the 2009 *Manual of Uniform Traffic Control Devices* (MUTCD) (*2*). The inclusion of the FYA in the MUTCD has led to widespread use throughout the United States. Subsequent field and simulation studies have shown that the FYA permissive indication is well understood by drivers (*3–5*). The Virginia Department of Transportation (VDOT) has adopted FYA signal displays as their preferred indication for permissive left-turns, including the permissive portion of protected-permissive left-turn (PPLT) phasing. With Virginia and other states adopting FYA and converting existing circular green permissive signal heads to FYA displays, the safety effectiveness of these conversions must be quantified.

Bayesian methods are currently the preferred approach for quantifying safety effectiveness of treatments because they are known to properly account for the regression-to-mean bias—a phenomenon which may yield artificially high estimates of the safety effectiveness of a treatment, especially if sites have been chosen for study based on their abnormally high crash frequencies (*6–8*). The Bayesian approach accounts for the regression-to-mean bias by combining prior information

[1]Virginia Transportation Research Council, Charlottesville, VA

Corresponding Author:
Address correspondence to Justice Appiah:
Justice.Appiah@VDOT.Virginia.Gov

with current information. In the context of a before–after safety analysis, the prior information would be the expected crash frequency from similar but untreated sites, and the current information would be crash frequency observed at the treated sites.

The empirical Bayes (EB) and full Bayes (FB) methods are two related approaches for combining prior information and current information. The EB method, currently the industry standard for conducting unbiased before–after evaluations, allows for out-of-sample estimation of the prior information by using data from a reference group of similar sites to develop a safety performance function (SPF) that relates crash frequencies to relevant site characteristics (*7*). The calibrated SPF is used to obtain point estimates of the expected crash frequency and its variance. This estimate of crash frequency is then combined with the crash frequency observed at a treatment site to obtain an improved estimate of the site's long-term expected crash frequency, had the treatment not been implemented. The long-term expected crash frequency estimate is compared with the actual crash frequency observed after the treatment was applied to quantify the treatment's effect on safety.

The FB approach also uses data from a reference group of intersections. However, the prior information is not estimated out-of-sample. Instead, data from the reference group, as well as data from the treated sites before the treatment was applied, are used in an integrated approach to obtain a distribution of estimates of the long-term expected crash frequency using Markov chain Monte Carlo methods (*9*, *10*). That is, instead of using a point estimate of the expected crash frequency and its variance, the FB approach generates a distribution of likely values that are then combined with site-specific crash data to obtain an estimate of the long-term expected crash frequency.

Though a relatively complex alternative to the EB approach, the use of a distribution rather than a point estimate enables relevant parameters such as the expected long-term crash frequency, the variance of the long-term crash frequency, and the variance of the estimated crash modification factors (CMFs) to be calculated more accurately (*8*). Other strengths of the FB approach relative to the EB approach include (i) more flexibility in selecting crash frequency distributions and model specification, (ii) the potential for estimating valid crash models with small sample sizes, (iii) the ability to incorporate spatial correlation between sites, (iv) the ability to incorporate prior knowledge of the values of model coefficients, and (v) the potential for more detailed causal inferences (*8*, *9*, *11*, *12*).

Objectives and Scope

As of late 2016, FYA was in use at more than 200 VDOT-maintained signalized intersections, with further installations planned at more than 150 other intersections. Anecdotal evidence suggests that these FYA installations have been effective at improving intersection safety. However, there has been no well-documented comprehensive analysis of their safety effectiveness in the state. The purpose of this research was to address this evaluative need as it pertains to FYA use in PPLT phasing. A secondary objective was to compare estimates of safety effectiveness obtained using EB with those obtained using FB. Although the FB approach offers several conceptual advantages, its relative complexity may act as a deterrent to agencies adopting the method for their evaluations. This study used FYA data as a case study to contrast results obtained by the two methods.

Crash records from before and after the implementation of FYA with PPLT phasing are compared. In addition, data from a group of similar but untreated intersections are used to compare crashes over time using Bayesian techniques to ensure that no exogenous variables affect the results in the long term.

Literature Review

This section summarizes the literature on the safety effects of using the FYA signal indication for the permissive portion of PPLT phasing. The review is specific to studies that evaluated conversions from PPLT with a circular green indication for the permissive phase to PPLT with FYA indication for the permissive phase.

Noyce et al. evaluated the effect of FYA on safety using several methodologies, including an EB before–after study for 19 treated intersections (*13*). This study was designed as a follow-up to the 2003 NCHRP study that recommended the use of FYA. No specific CMFs were provided partly because of a lack of sufficient "after" data. Nevertheless, the study provided some general insights into the safety effectiveness of FYAs. Overall, crashes tended to decrease at intersections that operated in the PPLT mode both before and after treatment, with the FYA replacing the permissive circular green indication (i.e., PPLT to PPLT–FYA conversion).

Srinivasan et al. evaluated the use of the FYA indication for permitted left-turns based on data from 51 signalized intersections in Oregon, Washington, and North Carolina (*14*). Analysis of the North Carolina data (16 intersections) was based on the EB method; that of Oregon and Washington was based on a combination of some aspects of the EB and the comparison group methods. The study found that total crashes and left-turn crashes decreased by 8% and 19%, respectively, for intersections that had PPLT to PPLT–FYA conversions. By comparing the standard deviation of the distribution of CMFs with the conventionally estimated standard error of the mean CMF, the authors found substantial variation in the effect of FYA treatment across the different sites and emphasized the importance of giving explicit consideration to CMF variability across sites when evaluating the effectiveness of safety treatments.

Qi et al. also investigated safety effectiveness of PPLT to PPLT–FYA conversions at 17 intersections in Tyler, TX and Kennewick, WA (*15*). For each intersection, the study estimated an odds ratio by comparing the left-turn crash rate observed after FYA installation to the EB estimate of the crash rate that would have occurred had the FYA installations not been made. The odds ratio estimates ranged from 0.00 to 4.55. Overall, the study concluded that the FYA treatment did not adversely affect safety at a majority of the study intersections; the odds ratio was greater than one at only three of the intersections.

Pulugurtha and Khader evaluated the safety effectiveness of FYAs at 18 intersections in Charlotte, NC using the EB method (*16*). Before-period conditions on the left-turn approaches treated included protected-only, permitted-only, PPLT, and stop-sign control. Individual intersection CMFs for left-turn crashes ranged between 0.00 and 1.52 with an average of 0.39, reflecting a 61% reduction in left-turn crashes. Five of the intersections had PPLT for left-turns prior to FYA installation. Crashes increased by 41% at one of these intersections and decreased by 49% to 100% at the other four. No standard error or confidence interval estimates were reported, so it is difficult to judge if these results are statistically significant. The authors recommended further studies using larger samples.

Simpson and Troy developed CMFs for FYA installations based on data from 222 intersections in North Carolina (*17*). CMFs were provided for total crashes, injury crashes, and target crashes (defined by the authors as "left-turn same-roadway crashes with the left-turner on an approach treated with FYA"). A simple before–after analysis method was used to compare crash frequencies before and after FYA treatment. SPFs from the *Highway Safety Manual* (HSM) (*7*) were used to account for the effects of changes in traffic volumes. The simple before–after method used in this study is not as robust as the EB method (*6*, *8*). Nevertheless, the study provides significant insights into the safety effectiveness of FYA treatment. For intersections that converted from PPLT to PPLT–FYA (105 total), the study found a 7% reduction in total crashes, a 15% reduction in injury crashes, and a 22% reduction in target crashes.

The current study adds to the literature by determining the effects of the PPLT–FYA installations on angle crash types. This may provide practitioners with a more practical safety factor than left-turn crashes as data on angle crashes are more readily available, easier to retrieve, and perhaps more relevant to practitioners than the movement-specific crash types. The Srinivasan et al. study also highlighted the need to explicitly consider CMF variability across sites when evaluating safety improvements (*14*). The FB method provides the flexibility to account for this uncertainty in an integrated approach. The use of both the EB and FB methods for determining safety effectiveness of PPLT–FYA also contributes to the discussion as to whether the benefits of the FB method outweigh the increased complexity (*8*).

Data

The study used data from 28 treated signalized intersections (see Figure 1) and a reference group of 39 signalized intersections across Virginia. At least one approach on each treated intersection had a left-turn that operated in the PPLT–FYA mode; all 3-legged intersections had one, and all the 4-legged intersections had two left-turns that operated in the PPLT–FYA mode. The treated approaches on all 28 intersections operated in the PPLT mode with circular green indication for the permissive portion. The number of sites analyzed was limited to 28 signalized intersections with PPLT–FYA because of factors such as traffic data availability, lack of installation date information, and close proximity to commercial driveways. It is noted that data were also available for eight other intersections that operated in the protected-only left-turn mode before being converted to PPLT–FYA. These protected-only to PPLT–FYA intersections were excluded from the analysis because of the small sample size and potential confounding effects. In particular for these intersections, any changes in crashes following PPLT–FYA conversion likely result from the introduction of the permissive phase and not necessarily the FYA signal indication, which was the focus of this paper. The reference group intersections did not have FYA signals installed at the time of the study; instead, these intersections were under consideration as candidates for future FYA installation by VDOT staff.

For each treated intersection, crash data and annual average daily traffic (AADT) data were retrieved from VDOT Traffic Engineering Division databases for the period from January 2008 through December 2016. The lengths of the before and after periods varied based on the date FYA was activated. Crash data were aggregated on a monthly basis, and for the FB analysis average daily traffic (ADT) estimates for each month of the analysis period were obtained by applying appropriate monthly adjustment factors generated by VDOT to the AADT data. To reduce possible novelty effects, data for the month in which the FYA signals were activated at an intersection were not used in the analysis. A summary of some relevant characteristics of the treated intersections is provided in Table 1. Similar data were also collected

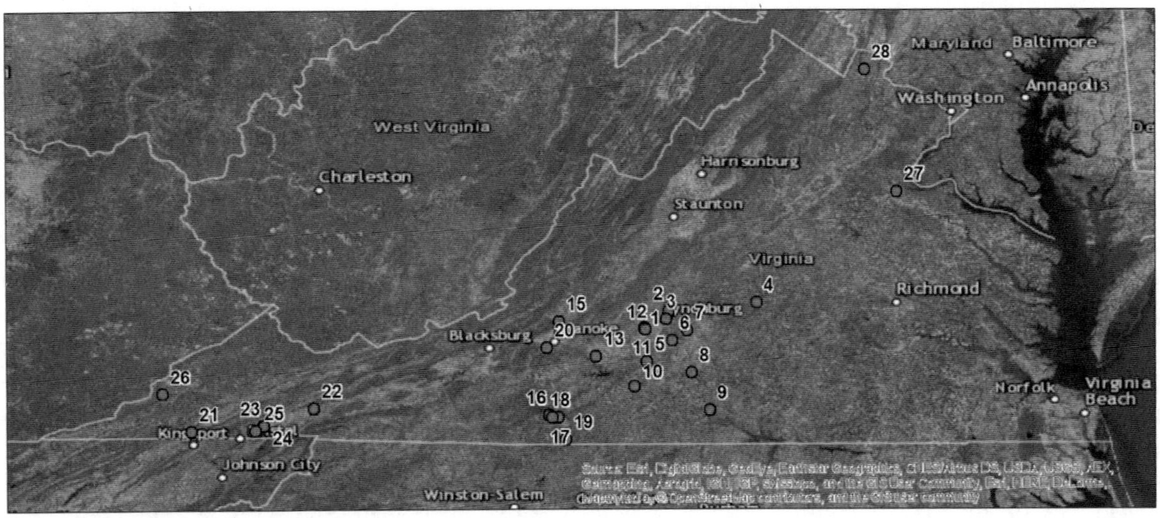

Figure 1. Map showing locations of treated intersections.

Table 1. Summary Data for Treated Intersections

Intersection ID	Number of legs	Approaches with FYA	Years		Main road AADT		Minor road AADT		Total crashes/year	
			Before	After	Before	After	Before	After	Before	After
1	4	2	7.0	1.9	19,702	22,613	7,449	7,350	10.3	1
2	3	1	7.0	1.9	13,542	13,917	1,422	1,467	2.0	2
3	3	1	7.0	1.9	9,940	9,357	2,119	2,090	1.4	3
4	4	2	7.0	1.9	8,887	8,745	2,281	2,502	3.1	4
5	3	1	6.9	2.0	9,850	9,701	2,753	2,804	1.6	5
6	3	1	6.9	2.0	3,678	4,216	1,886	1,915	1.4	6
7	4	2	6.9	2.0	17,924	18,052	2,832	3,198	5.9	7
8	3	1	7.5	1.4	5,915	5,756	1,172	1,218	0.7	8
9	4	2	7.0	1.9	8,270	7,698	546	546	1.6	9
10	4	2	7.0	1.9	4,484	4,463	3,978	4,037	0.9	10
11	3	1	6.8	2.2	20,830	23,882	6,273	7,035	5.0	11
12	3	1	6.8	2.2	13,216	15,096	2,362	2,406	1.2	12
13	4	2	6.8	2.1	7,342	8,066	3,275	3,560	0.6	13
14	3	1	5.3	3.7	25,953	24,748	1,692	1,592	4.0	14
15	3	1	8.2	0.8	20,540	21,716	2,814	2,685	2.9	15
16	4	2	7.0	1.9	5,831	5,402	3,548	3,142	2.7	16
17	4	2	7.0	1.9	13,572	12,444	3,942	3,524	3.6	17
18	3	1	7.0	1.9	7,214	6,803	4,009	3,496	1.7	18
19	3	1	7.0	1.9	14,305	14,988	3,894	2,968	7.1	19
20	4	2	5.3	3.7	28,605	27,956	8,625	8,832	17.5	20
21	3	1	6.8	2.1	14,321	13,032	5,642	5,697	2.2	21
22	4	2	7.1	1.8	8,483	8,361	4,472	4,303	1.7	22
23	3	1	7.2	1.8	9,798	8,627	994	948	1.7	23
24	4	2	6.8	2.1	8,635	7,720	3,282	3,760	4.4	24
25	3	1	7.2	1.8	9,165	8,402	1,296	894	1.1	25
26	3	1	7.2	1.8	6,915	5,148	1,285	1,114	0.6	26
27	3	1	5.2	3.8	10,268	10,786	4,605	5,163	5.4	27
28	4	2	4.6	4.3	15,156	16,068	6,272	7,631	7.4	28
Mean			6.8	2.2	12,226	12,277	3,383	3,424	3.6	3.3

Note: AADTs and crashes are average annual values for the before and after years.

for the reference group of intersections. In retrieving the crash data, a crash event was assigned to an intersection if it was within 250 feet of the intersection, which was the distance utilized in the development of past Virginia DOT SPFs (*18*). The major road was defined as the road with the higher AADT.

Methods

The safety effectiveness of PPLT–FYA installations in Virginia were evaluated using both the FB approach and the EB approach. The reference list of 39 intersections was considered adequate for the FB approach but not large enough to estimate a reliable out-of-sample crash reduction model required for the EB method. Consequently, a precalibrated Virginia-specific intersection SPF was used for the EB analysis.

Empirical Bayes Method

When using the EB method, SPFs are used to predict the number of crashes at a location. As all sites were in Virginia, Virginia-specific SPFs were used in the calculations. For Virginia signalized intersections, prior research developed 32 SPFs (*18*) which are currently used by VDOT. These SPFs were developed to cover the following factors:

- Statewide or regional SPFs
- Urban or rural location
- 3 or 4 legs
- Total crashes (all types and severities) or fatal and injury crashes (all types)

Regional SPFs were developed for different parts of the state; each was deemed to have common geometric and driver behavior characteristics. The Northern Region was composed of primarily urban and suburban counties near Washington, D.C. that often experienced higher levels of traffic and congestion than the other parts of the state. The Western Region consisted of primarily rural roads that often were located in rolling or mountainous terrain. The Eastern region consisted of counties in the central and eastern part of the state that were primarily flat and had a mixture of urban and rural counties. More precise descriptions of the regions and the model development process can be found elsewhere (*18*). Each of the intersections was analyzed using the corresponding SPF for their region, location type, and intersection geometry. In addition, separate SPFs were used for total crashes and fatal and injury crashes. The form of the equation for the Virginia-specific SPFs is shown in Equation 1, and the parameter values are summarized in Table 2:

$$N = e^{\alpha} \times (Major\,AADT)^{\beta_1} \times (Minor\,AADT)^{\beta_2} \quad (1)$$

Table 2. SPF Parameters

Region	Crash severity	Urban/rural	Intersection legs	α	β_1	β_2	k
Northern	All	Urban	3	−4.999	0.556	0.155	0.393
		Urban	4	−8.307	0.752	0.328	0.212
		Rural	3	−5.880	0.508	0.313	0.391
		Rural	4	−1.604	0.228	0.151	0.321
	Fatal and injury	Urban	3	−7.398	0.650	0.209	0.415
		Urban	4	−9.655	0.760	0.360	0.206
		Rural	3	−9.219	0.726	0.341	0.735
		Rural	4	−3.329	0.360	0.060	0.251
Western	All	Urban	3	−9.614	0.868	0.330	0.372
		Urban	4	−12.391	1.063	0.457	0.162
		Rural	3	−6.437	0.544	0.286	0.411
		Rural	4	−6.395	0.551	0.311	0.153
	Fatal and injury	Urban	3	−11.010	0.908	0.323	0.504
		Urban	4	−11.428	0.866	0.441	0.149
		Rural	3	−8.861	0.706	0.281	0.392
		Rural	4	−8.058	0.681	0.256	0.229
Eastern	All	Urban	3	−6.752	0.616	0.297	0.334
		Urban	4	−8.855	0.783	0.371	0.135
		Rural	3	−9.899	0.805	0.439	0.495
		Rural	4	−0.647	0.226	−0.016	0.391
	Fatal and injury	Urban	3	−7.266	0.551	0.311	0.298
		Urban	4	−9.958	0.748	0.402	0.127
		Rural	3	−11.184	0.867	0.398	0.106
		Rural	4	−4.838	0.424	0.152	0.329

Note: *k* = over-dispersion factor.

The base number of predicted crashes (N) determined with the SPFs was adjusted for site-specific conditions by use of appropriate CMFs presented in the HSM (*7*). The EB method combines the observed and predicted number of crashes to form an expected number of crashes for the before-installation period for each site. This number, along with the ratio of the predicted number of crashes in the after period to the predicted number of crashes in the before period, is then used to calculate the number of expected crashes in the after period. From there, this value is compared with the observed number of crashes at each site to find an odds ratio (OR). The overall OR is then calculated using the summation of all observed crashes at all locations and dividing that by the summation of all predicted crashes at all locations. Finally, a correction is applied to this value to formulate the final CMF value for the analysis. All calculations were done using standard HSM procedures (*7*).

Traditionally, the installation-year data would not be used for CMF calculations, which would eliminate sites that did not have a full year's worth of after data. To allow for a larger sample size and hopefully more accurate results than analyzing only sites with full-year data, sites with partial-year data were included. For these sites, the lack of a full year's worth of data was accounted for by multiplying the number of predicted crashes in both the before and after periods by the number of months before or after and dividing by 12. For example, for a site that was installed in March of 2016, the SPF-predicted crashes for 2016 in the before period was multiplied by 2/12, and the number of SPF-predicted crashes for 2016 in the after period was multiplied by 9/12. In addition, the Virginia SPFs used for the EB analysis were developed for total crashes and fatal and injury crashes; they do not explicitly model angle crashes or other crash types. To correct for only angle crash types, the number of observed crashes for each year only counted angle crashes, and the number of predicted crashes for both the before and after periods were multiplied by the percentage of angle crashes to total crashes before the conversion.

Full Bayes Method

Two basic model forms that account for over-dispersion in crash counts were considered for the FB analysis: the Poisson-Gamma model and the Poisson-Lognormal model (*9*, *12*, *19*). These models can be expressed as:

$$y_{it} \sim Pois(\varepsilon_i \lambda_{it}) \tag{2}$$

where

y_{it} = observed number of crashes at intersection i in month t

ε_i = nonnegative multiplicative random effect term for intersection i (to model individual heterogeneity)

For the Poisson-Gamma model,

$\varepsilon_i \sim Gamma(\phi, 1/\phi)$ with mean 1, variance $1/\phi$ and dispersion parameter φ

$\varepsilon_i \sim Gamma(1, 1)$

For the Poisson-Lognormal model,

$\varepsilon_i \sim LogN(0, \sigma^2)$ with mean $e^{\sigma^2/2}$ and variance $e^{\sigma^2}(e^{\sigma^2} - 1)$

$\sigma^2 \sim Inverse\ Gamma(0.001, 0.001)$

λ_{it} = expected number of crashes at intersection i in month t

The expected number of crashes (λ_{it}) was modeled as a function of entering ADTs on the major and the minor roads. Three functional forms of the regression model were tested; they are shown in Equations 3–5. Equation 4 includes a time trend term to account for time effects while Equations 3 and 5 do not. Equation 5 also includes time-varying coefficients to account for temporal variations in traffic environment, driver behavior, weather, and so forth (*9*, *11*, *12*):

$$\lambda_{it} = d_{3i}\beta_{30} + d_{4i}\beta_{40} + (d_{3i}\beta_{31} + d_{4i}\beta_{41})\ln\left(\frac{ADTm_{it}}{1000}\right) + (d_{3i}\beta_{32} + d_{4i}\beta_{42})\ln\left(\frac{ADTc_{it}}{1000}\right) \tag{3}$$

$$\lambda_{it} = d_{3i}\beta_{30} + d_{4i}\beta_{40} + (d_{3i}\beta_{31} + d_{4i}\beta_{41})\ln\left(\frac{ADTm_{it}}{1000}\right) + (d_{3i}\beta_{32} + d_{4i}\beta_{42})\ln\left(\frac{ADTc_{it}}{1000}\right) + \beta_3 t \tag{4}$$

$$\lambda_{it} = d_{3i}\beta_{t30} + d_{4i}\beta_{t40} + (d_{3i}\beta_{31} + d_{4i}\beta_{41})\ln\left(\frac{ADTm_{it}}{1000}\right) + (d_{3i}\beta_{32} + d_{4i}\beta_{42})\ln\left(\frac{ADTc_{it}}{1000}\right) \tag{5}$$

where

β_{kj} (k = 3, 4; j = 0, 1, 2), β_{t30}, β_{t40}, and β_3 are regression coefficients

d_{ki} (k = 3, 4) are dummy variables such that d_{ki} = 1 if intersection i is k-legged

otherwise,

$ADTm_{it}$ = main road entering ADT for intersection i and month t

$ADTc_{it}$ = minor road entering ADT for intersection i and month t

Markov chain Monte Carlo methods were used to calibrate posterior distributions for the model parameters and functions of model parameters using the WinBUGS software (*10*). The calibration was done using data from the period before treatment occurred at the treated sites and data from the entire before and after

Table 3. Summary Results for EB Analysis

	Crash reduction rate (R)	Crash modification factor (CMF)	Standard error
Total crashes	0.00	1.00	0.079
Fatal and injury crashes	−0.07	1.07	0.135
Angle crashes	0.19	0.81*	0.101

*Indicates CMF is statistically significant at the 90% confidence level.

Table 4. Deviance Information Criterion Estimates

Model	Total crashes	Fatal and injury crashes	Angle crashes
Poisson-Gamma; fixed intercept; no time trends	9110	5097	4593
Poisson-Gamma; fixed intercept; time trends	9108	5096	4598
Poisson-Gamma; time-varying intercept; no time trends	9279	5278	4778
Poisson-Lognormal; fixed intercept; no time trends	9149	5136	4618
Poisson-Lognormal; fixed intercept; time trends	9145	5136	4618
Poisson-Lognormal; time-varying intercept; no time trends	9279	5276	4771

periods for the reference sites. Prior distributions for all regression coefficients were assumed normal with mean zero and a large variance (10^3) to reflect complete ignorance about their values, a priori. The calibrated parameters were used to obtain estimates of the total number of crashes expected ($\varepsilon_i \lambda_{it}$) at the treated sites in the after periods had the treatment not been implemented. The expected crash reduction rate (R) following FYA installation was then calculated as the percentage difference between the expected total number of crashes and the actual number of crashes observed in the after period:

$$R = 1 - \frac{\sum_{i \in I} \sum_{t > T_{0i}} y_{it}}{\sum_{i \in I} \sum_{t > T_{0i}} \varepsilon_i \lambda_{it}} \quad (6)$$

where,

T_{0i} = month in which FYA was activated at intersection I

I = set of treated intersections

y_{it} = observed number of crashes at intersection i in month t

Analysis

Empirical Bayes Analysis and Results

Separate analyses were performed for:

- Total crashes (all types and severities)
- Fatal and injury crashes (all types)
- Angle crashes (all severities)

CMFs and their corresponding standard errors and significance levels were computed and are shown in Table 3. As shown in Table 3, the PPLT to FYA–PPLT conversion created 19% reduction in angle crashes (all severities), which was statistically significant at the 90% confidence level. Other changes were not statistically significant.

Full Bayes Analysis and Results

Model parameters were treated as random variables and the goal was to estimate the posterior distribution of likely values of the parameters given the data and prior information (*20*). Two parallel Markov chains were run for 130,000 iterations to obtain posterior distributions of the model parameters and the crash reduction rates. Convergence was monitored using the Brooks–Gelman–Rubin diagnostic (*10*), which showed that model parameters converged after approximately 30,000 iterations. Consequently, the first 30,000 iterations of each chain were discarded as burn-in runs. In addition, the chains were thinned using a factor of 20 (i.e., results were collected on model parameters at every 20th iteration only) so as to minimize the effects of serial correlations; thus, inferences were based on samples of size 10,000 for every model parameter.

The deviance information criterion (DIC), a goodness-of-fit measure similar to Akaike's information criterion, was used as the primary criterion for comparing the different FB models. The DIC considers the posterior expectation of the deviance as a measure of fit and includes an estimate of the effective number of parameters as a penalty measure for the complexity of the

Table 5. Posterior Moments and Quantiles of Model Parameters

Node	Mean	Standard error	Median	95% credible interval	
				2.50%	97.50%
Total crashes					
β_{30}	−4.945	0.969	−5.006	−6.660	−2.854
β_{40}	−6.198	0.703	−6.230	−7.498	−4.716
β_{31}	0.295	0.235	0.299	−0.184	0.737
β_{41}	0.559	0.146	0.559	0.266	0.848
β_{32}	0.515	0.172	0.518	0.163	0.840
β_{42}	0.329	0.101	0.329	0.132	0.525
β_3	0.002	0.001	0.002	0.000	0.003
R	0.119	0.053	0.122	0.010	0.216
Fatal and injury crashes					
β_{30}	−5.773	1.171	−5.828	−7.876	−3.254
β_{40}	−7.667	0.738	−7.705	−9.014	−6.068
β_{31}	0.525	0.300	0.532	−0.088	1.093
β_{41}	0.746	0.158	0.750	0.423	1.048
β_{32}	0.502	0.228	0.506	0.046	0.939
β_{42}	0.272	0.116	0.271	0.044	0.501
β_3	0.001	0.001	0.001	−0.001	0.004
R	0.137	0.079	0.140	−0.032	0.283
Angle crashes					
β_{30}	−6.390	1.308	−6.439	−8.771	−3.743
β_{40}	−5.685	0.804	−5.700	−7.194	−4.029
β_{31}	0.741	0.361	0.739	0.021	1.465
β_{41}	0.600	0.219	0.604	0.167	1.019
β_{32}	0.836	0.251	0.841	0.335	1.314
β_{42}	0.083	0.146	0.085	−0.208	0.365
β_3	0.002	0.001	0.002	−0.001	0.004
R	0.301	0.066	0.304	0.162	0.421

model. In comparing two models, the model with the lower DIC is generally preferable (*10*). DIC estimates for the different models tested are summarized in Table 4. The Poisson-gamma model with a regression function for the expected number of crashes that includes a time trend term to allow for estimation and accounting for time effects was selected because it had relatively low DIC values and the 90% posterior credible interval for the trend effect did not include zero.

Separate models were calibrated based on data for:

- Total crashes (all types and severities)
- Fatal and injury crashes (all types)
- Angle crashes (all severities)

In each case, posterior distributions of the crash reduction rates were monitored for all treated intersections. The results of the FB analysis are summarized in Table 5.

The results indicated a fairly high reduction of 30% in angle crashes. Reductions in total crashes (12%) and fatal and injury crashes (14%) were relatively modest. Although the average reduction in fatal and injury crashes was fairly substantial, the study did not rule out potential increases in fatal and injury crashes following conversion from PPLT to FYA–PPLT. The 95% credible interval for the fatal and injury crash reduction rate included zero (no safety effects) and negative values (increased crashes). However, the results also suggest that fatal and injury crash reduction rates as high as 28% are probable with this FYA treatment type.

Discussion

In reviewing the results presented above, a reliable CMF was calculated using the EB method for angle crashes (all severities); CMF estimates for total crashes (all types and severities) and fatal and injury crashes (all types) were not statistically significant.

Using the FB approach, reliable CMFs (calculated as, 1-*R*) were obtained for total crashes and angle crashes (all severities). The CMF for fatal and injury crashes (all types) was 0.86, which is similar to the CMF of 0.85 obtained by Simpson and Troy (*17*). However, the central 95% credible interval did not rule out the possibility of CMFs greater than one (increased crashes). Also, it is worth noting that the angle crashes CMF was very high compared with the total crashes CMF. This shows that

Table 6. Comparing Crash Reduction Rates (*R*) Estimated by the EB and FB Approaches

	Empirical Bayes	Full Bayes
Total crashes (all types and all severities)	0.00 (0.079)	0.12 (0.053)
Fatal & injury crashes (all types)	−0.07 (0.135)	0.14 (0.079)
Angle crashes (all severities)	0.19 (0.101)	0.30 (0.066)

Note: Standard errors are in parentheses.

PPLT–FYA conversion had a more significant impact on angle crashes, which is to be expected.

For purposes of comparison, the results for different crash types and severities using the FB and EB methods are shown in Table 6. First, it can be seen that the FB approach can provide similar results to the EB; the statistically significant angle crash reduction rate estimated by the EB was within the central 95% credible set of the corresponding FB estimate. Second, even with the relatively small sample of conversion sites, the 95% credible sets for expected crash reduction rates estimated with the FB method are relatively narrow as compared with EB results, suggesting that CMFs are estimated with a good degree of confidence. This is consistent with the belief that the FB can have an advantage over the EB when the sample size is restricted because of cost and other practical limitations (*8*, *11*, *12*).

The estimated 12% (1–22%) reduction in total crashes for the PPLT to PPLT–FYA conversion is consistent with the 12% and 6% estimates reported by Srinivasan et al. and Simpson and Troy, respectively (*14*, *17*). Srinivasan et al. also reported a 19% decrease in left-turn crashes following conversion from PPLT to PPLT–FYA. Left-turn crashes were not studied as part of this research; instead, angle crashes were analyzed because information on angle crashes is more readily accessible by VDOT engineers. It is worth noting that the expected decrease in left-turn crashes of 19% obtained in the Srinivasan et al. study and the expected decrease in target crashes (crashes involving left-turning vehicles and opposing through vehicles) of 22% reported in the Simpson and Troy study are both consistent with the 30% (16–42%) expected decrease in angle crashes obtained in this study. The expected reduction in fatal and injury crashes of 14% (–3% to 28%) following FYA conversion is also consistent with the 15% reduction reported by Simpson and Troy (*17*).

Conclusion

This paper examined results of the safety changes attained as a result of implementing FYA signals at 28 signalized intersections in Virginia using standard Bayesian methods. The results indicated that converting left-turns from PPLT to PPPLT–FYA tend to improve intersection safety, which seems to support the findings of previous studies that drivers understand the FYA signal indication better than the circular green. For the intersections studied in this research, the expected crash reduction following conversion from PPLT to PPLT–FYA was estimated as 12% (total crashes), 14% (fatal and injury crashes), and 30% (angle crashes).

The results also suggested that the FB approach can provide similar results to the well-established EB approach to safety evaluation. The 95% credible sets for expected crash reduction rates estimated with the FB method were generally narrow, suggesting a good degree of confidence in the estimates. This is consistent with the findings in other studies that have suggested that the FB can have an advantage over the EB when the sample size is restricted because of cost and other practical limitations (*12*).

The largest limitation of the EB analysis was the number of sites found in Virginia that underwent a conversion from PPLT to PPLT–FYA. Another limitation was the lack of a Virginia-specific SPF for angle crashes. It would be useful in the future, therefore, to conduct a similar comparative evaluation of EB and FB when more data become available. Any such comparison study will benefit from the use of a well-developed SPF for angle crashes. Finally, all analyses were based on having two treated approaches at four-legged intersections. Additional research is needed to assess effects when three or all four approaches are treated.

Acknowledgments

The authors gratefully acknowledge the contributions of numerous individuals who made the successful completion of this paper possible. They are especially grateful to Michael Clements, Don Robbins, Sean Becker, and In-Kyu Lim of VDOT's Traffic Engineering Division. The contributions of Ramkumar Venkatanarayana of the Virginia Transportation Research Council are also appreciated.

Author Contributions

The authors confirm contribution to the paper as follows: study conception and design: Appiah, Fontaine, Cottrell, King; data collection: Appiah, King, Fontaine, Cottrell; analysis and interpretation of results: Appiah, King, Fontaine, Cottrell; draft manuscript preparation: Appiah, King, Fontaine, Cottrell. All authors reviewed the results and approved the final version of the manuscript.

References

1. Brehmer, C. L., K. C. Kacir, D. A. Noyce, and M. P. Manser. *NCHRP Report 493: Evaluation of Traffic Signal Displays for Protected/Permissive Left-turn Control.* Transportation Research Board of the National Academies, Washington, D.C., 2003.
2. *Manual on Uniform Traffic Control Devices for Streets and Highways.* FHWA, U.S. Department of Transportation, Washington, D.C., 2009.
3. Knodler, M. Jr., D. Noyce, and D. Fisher. Evaluating Effect of Two Allowable Permissive Left-Turn Indications. *Transportation Research Record: Journal of the Transportation Research Board,* 2007. 2018: 53–62.
4. Rietgraf, A., and K. Schattler. Behavior of Left-Turning Drivers during Permissive Interval of Protected-Permissive Operation: Effect of Signal Display. *Transportation Research Record: Journal of the Transportation Research Board,* 2013. 2384: 35–44.
5. Lin, P. W., G. Thiagarajan, and D. Atie. Analysis of Drivers' Reaction to the Flashing Yellow Arrow Signal Design from Field Observation. Presented at 15th World Congress on Intelligent Transport Systems and ITS America's 2008 Annual Meeting, New York, NY, 2008.
6. Hauer, E. *Observational Before-After Studies in Road Safety.* Elsevier Science Inc., New York, NY, 1997.
7. *Highway Safety Manual.* AASHTO, Washington, D.C., 2010.
8. Gross, F., B. Persaud, and C. Lyon. *A Guide to Developing Quality Crash Modification Factors.* FHWA-SA-10-032. FHWA, U.S. Department of Transportation, Washington, D.C., 2010.
9. Lan, B., B. Persaud, C. Lyon, and R. Bhim. Validation of a Full Bayes Methodology for Observational Before-After Road Safety Studies and Application to Evaluation of Rural Signal Conversions. *Accident Analysis and Prevention,* Vol. 41, No. 3, 2009, pp. 574–580.
10. Spiegelhalter, D., A. Thomas, N. Best, and D. Lunn. *WinBUGS Version 1.4 User Manual.* MRC Biostatistics Unit, Cambridge, MA, 2003.
11. Carriquiry, A. L., and M. Pawlovich. *From Empirical Bayes to Full Bayes: Methods for Analyzing Traffic Safety Data.* White Paper. Iowa State University, Ames, IA, 2004.
12. Persaud, B., B. Lan, C. Lyon, and R. Bhim. Comparison of Empirical Bayes and Full Bayes Approaches for Before–After Road Safety Evaluations. *Accident Analysis & Prevention* Vol. 42, No. 1, 2010, pp. 38–43.
13. Noyce, D., C. Bergh, and J. Chapman. *NCHRP Report W123: Evaluation of the Flashing Yellow Arrow Permissive-Only Left Turn Indication Field Implementation.* Transportation Research Board of the National Academies, Washington, D.C., 2007.
14. Srinivasan, R., C. Lyon, B. Persaud, J. Baek, F. Gross, S. Smith, and C. Sundstrom. Crash Modification Factors for Changes to Left-Turn Phasing. *Transportation Research Record: Journal of the Transportation Research Board,* 2012. 2279: 108–117.
15. Qi, Y., M. Zhang, Y. Wang, and X. Chen. Safety Performance of Flashing Yellow Arrow Signal Indication. Presented at 91st Annual Meeting of the Transportation Research Board, Washington, D.C., 2012.
16. Pulugurtha, S., and K. Khader. Assessing the Effect of Introducing a Permitted Phase Through the Use of Flashing Yellow Arrow Signal for Left-Turning Vehicles. *International Journal of Injury Control and Safety Promotion,* Vol. 21, No. 4, 2014, pp. 338–337.
17. Simpson, C. L., and S. A. Troy. Safety Effectiveness of Flashing Yellow Arrow: Evaluation of 222 Signalized Intersections in North Carolina. *Transportation Research Record: Journal of the Transportation Research Board,* 2015. 2492: 46–56.
18. Garber, N. J., and G. Rivera. *Safety Performance Functions for Intersections on Highways Maintained by the Virginia Department of Transportation.* VTRC 11-CR1. Virginia Transportation Research Council, Charlottesville, VA, 2010.
19. Zhou, M., L. Li, D. Dunson, and L. Carin. Lognormal and Gamma Mixed Negative Binomial Regression. In Langford, J., and J. Pineau (eds.), *Proc., International Conference on Machine Learning,* Omnipress, Madison, WI, 2012, pp. 1343–1350.
20. Pawlovich, M. D., W. Li, A. Carriquiry, and T. Welch. Iowa's Experience with Road Diet Measures: Use of Bayesian Approach to Assess Impacts on Crash Frequencies and Crash Rates. *Transportation Research Record: Journal of the Transportation Research Board,* 2006. 1953: 163–171.

The Standing Committee on Traffic Control Devices (AHB50) peer-reviewed this paper (18-01468) and selected it to receive its 2018 Best Paper Award.

Article

TRR
JOURNAL OF THE TRANSPORTATION RESEARCH BOARD

Transportation Research Record
2018, Vol. 2672(21) 30–40

DOI: 10.1177/0361198118776481
journals.sagepub.com/home/trr
SAGE

Safety Evaluation of Alternative Audible Lane Departure Warning Treatments in Reducing Traffic Crashes: An Empirical Bayes Observational Before–After Study

Lingtao Wu[1], Srinivas R. Geedipally[2], and Adam M. Pike[1]

Abstract

Roadway departure crashes are a major contributor to traffic fatalities and injury. Rumble strips have been shown to be an effective countermeasure in reducing roadway departure crashes. However, some roadway situations, for instance, inadequate shoulder width or roadway surface depth, have limited the application of conventional milled or rolled in rumble strips. Alternative audible lane departure warning systems, including profile (audible) pavement markings and preformed rumble bars, are increasingly used to overcome the limitations that exist with the milled rumble strips. So far, the safety effectiveness of these alternative audible lane departure warning systems has not been extensively assessed. The main purpose of this paper is to examine the safety effect of installing profile pavement markings and preformed rumble bars. Specifically, this study developed crash modification factors for these treatments that quantify the effectiveness in reducing single-vehicle-run-off-road (SVROR) and opposite-direction (OD) crashes. Traffic, roadway, and crash data at the treated sites on 189 miles of rural two-lane highways in Texas were analyzed using an empirical Bayes (EB) before–after analysis method. Safety performance functions from the *Highway Safety Manual* and Texas Highway Safety Design Workbook were used in the EB analysis. The results revealed a 21.3% reduction in all SVROR and OD crashes, and 32.5% to 39.9% reduction in fatal and injury SVROR and OD crashes after installing profile pavement marking and preformed rumble bars.

Roadway departure crashes are a major contributor to highway fatalities in most states in the U.S.A. (*1*). From 2013 to 2015, an average of 18,275 fatalities per annum resulted from roadway departure crashes, which is 54% of all the traffic fatalities in the U.S.A. Single-vehicle-run-off-road (SVROR) and opposite-direction (OD) crashes are the two major outcomes of vehicles departing their lanes. SVROR crashes are the most prevalent among all crash types. About 18,000 fatal single-vehicle collisions occurred on U.S. roadways in 2016, resulting in approximately 11,700 fatalities and 814,000 injuries (*2*, *3*). Although OD crashes are less frequent compared to SVROR, these crash types are usually more severe. Many countermeasures have been implemented to reduce the number and/or severity of lane departure crashes, such as shoulder treatments (e.g., paved shoulder, widening shoulder), delineation improvements, rumble strips, markings, roadside and median barriers, etc. Among these treatments, rumble strips are frequently used for their low installation and maintenance costs, and effectiveness in reducing SVROR and OD crashes.

Rumble strips are usually placed longitudinal to the roadway surface on the shoulder, edgeline, or centerline of the pavement to alert drivers who have drifted out of their travel lanes. The auditory and tactile warnings produced by running over rumble strips help alert drivers to correct their path. Rumble strips are typically milled into the road surface, but require adequate road surface depth and shoulder space. Alternative treatments to milled rumble strips are becoming more common because of the safety benefits that milled rumble strips have continually shown. Alternative systems are typically profile (audible) pavement markings, but have recently included preformed rumble bars. These countermeasures are commonly used on seal coat (chip seal) road surfaces where

[1]Texas A&M Transportation Institute, Texas A&M University System, TAMU, College Station, TX
[2]Texas A&M Transportation Institute, Texas A&M University System, Arlington, TX

Corresponding Author:
Address correspondence to Lingtao Wu: wulingtao@gmail.com

Figure 1. Profile pavement markings (left) and preformed rumble bars (right) (*4*).

milled rumble strips cannot be utilized or on roadways where the shoulder is too narrow for milled rumble strips. Similar to rumble strips, profile pavement markings and rumble bars are intended to alert drivers that they are leaving their lanes through auditory and tactile warnings. The main difference is that rumble strips are typically milled or rolled into the pavement, while the profile pavement markings and rumble bars are usually installed on the markings. Figure 1 demonstrates an example of the profile pavement markings and preformed rumble bars. There is currently no guidance at the national level on when and where to use alternative rumble strip treatments. A few states have standards and specifications that outline specific layouts and designs of the treatments. More details about the design, performance, and characteristics of profile pavement markings and rumble bars are described in Pike et al. (*4*).

Although a considerable number of studies have reported the effectiveness of rumble strips, most of them focus on milled rumble strips. Little research has been conducted to determine if the alternative systems (i.e., profile pavement markings and rumble bars) provide a similar safety benefit to milled rumble strips. Thus, the main objective of this study is to evaluate the safety effectiveness of the profile pavement markings and rumble bars in reducing SVROR and OD crashes.

The rest of this paper is organized as follows. The second section provides a background pertaining to crash modification factors and safety effects of rumble strips. The third section documents the detailed methodology used in this study for estimating the safety effectiveness of profile pavement markings and rumble bars. The fourth section describes the dataset used in this study. The results are discussed in section five. The final section summarizes the study findings.

Background

This section first discusses the methods commonly used to assess the safety effectiveness of a treatment. The section then reviews the literature related to safety effectiveness of rumble strips.

Crash Modification Factors and Safety Evaluation Methods

In traffic safety studies, the effectiveness of a treatment is represented in the form of a crash modification factor (CMF). CMFs play a significant role in roadway safety management, including safety effect evaluation, crash prediction, hotspot identification, countermeasure selection, and economic evaluation. Several methods have been proposed for developing CMFs, such as before–after methods, cross-sectional studies (e.g., regression models and case control), expert panel studies, and propensity score framework (*5–8*). Among these methods, before–after studies are always preferred whenever available, since this approach is able to capture the cause–effect relationship and produces more reliable CMFs (*9, 10*). There are three common types of before–after studies for developing CMFs, i.e., naïve or simple before–after, before–after with comparison group, and empirical Bayes (EB) method. A naïve before–after study simply assumes that the number of crashes that occurred before the treatment is a good estimate of the number of crashes that may occur in the after period if the site had not been treated, without accounting for the regression-to-the-mean (RTM) bias and other changes. A before–after study with comparison group can overcome some of the limitations of naïve studies but it still cannot address RTM bias. It is generally believed that naïve and comparison group before–after studies are unable to generate reliable CMFs due to these limitations. The EB method has been recognized as the state-of-the-art approach for developing CMFs (*5, 11*). As a result, this study utilizes the EB method to estimate the CMFs of profile pavement markings in reducing lane departure crashes. Note that the lane departure warning systems in this study mean profile pavement markings and rumble bars (the two terms will be used interchangeably in the rest of this paper).

Safety Effectiveness of Rumble Strips

A considerable number of studies have evaluated the safety effectiveness of rumble strips. Torbic et al. (*12*) collected crash data from more than 600 miles of undivided two-lane highways from three states from 1997 through 2006. The researchers evaluated the safety effectiveness of installing milled centerline and edgeline rumble strips with the EB method. Results indicated that the installation of milled shoulder rumble strips on rural two-lane highways reduces SVROR crashes by 16%, and centerline rumble strips were effective in reducing head-on and OD sideswipe crashes on undivided rural two-lane highways.

Patel et al. (*13*) estimated the safety effectiveness of installing milled shoulder rumble strips on rural two-lane highways in Minnesota. Shoulder rumble strips were installed at 24 treatment sites (for a total length of 183 miles) on rural two-lane highways between 1995 and 2001. SVROR crash data were collected before and after the installation of the treatment. The researchers analyzed the data using the EB method. Results revealed that 13% of SVROR crashes were reduced after the treatment.

Sayed et al. (*14*) concluded that installing shoulder rumble strips on rural two-lane highways reduces SVROR crashes by 26.1%. Park and Abdel-Aty (*15*) assessed the safety effectiveness of installing rumble strips on rural two-lane highways in Florida using the before–after EB method. The results revealed that shoulder rumble strips reduce SVROR crashes by 25%. Rys et al. (*16*) conducted a before-after study to evaluate the safety effectiveness of milled centerline rumble strips in Kansas and concluded that head-on and OD sideswipe crashes decreased by 67.2%.

More recently, Khan et al. (*17*) examined the safety effectiveness of installing milled shoulder rumble strips in reducing SVROR crashes on rural two-lane highways. Shoulder rumble strips were installed at 38 sites on three highways in Idaho between 2004 and 2007. The researchers collected the roadway segment characteristics (e.g., length, average daily traffic (ADT), curvature, shoulder width, etc.) and utilized a before–after EB method. The results showed that SVROR crashes decreased by 14% after installing shoulder rumble strips on rural two-lane highways.

As can be seen, most of the previous studies focused on milled-in rumble strips. The number of studies on profile markings is very limited, probably because the majority of rumble strips are milled and rolled. Some studies focused on the performance of profile markings, such as retroreflectivity, cost, noise level, etc. (*18–21*). Among these studies, Lindly and Wijesundera (*20*) compared crash rates of flat thermoplastic markings and profiled pavement markings using a simple before–after method, but no significant safety benefit was found for profile pavement markings. Hatfield et al. (*22*) evaluated the safety effectiveness of installing profile markings on reducing crashes in Australia. The researchers collected data on about 107 miles of divided highways and undivided highways. Multiple years of crash data before and after treatment on these segments were analyzed using a before–after EB method. The analysis showed that installing profile edgeline markings on highways (combination of divided and undivided) would reduce passenger car SVROR crashes and truck SVROR crashes by 44% and 61%, respectively. It is unknown whether these results apply in the U.S. because of the differences in weather, roadway characteristics, and driver behavior.

Methodology

This section documents the approach used in this study to quantify the effectiveness of profile pavement markings and rumble bars. Specifically, the first part briefly describes the EB method, and the second part discusses the safety performance functions (SPFs) used in the EB approach.

EB Method

As mentioned above, before–after study with EB approach has been recognized as a robust method for developing CMFs. The EB method is able to account for the RTM bias and other changes over time not due to the treatment, and to reduce the level of uncertainty in the estimates of safety effect (*23*).

In most traffic safety before–after studies, the CMF is estimated by comparing the number of expected crashes that would have occurred in the after period had the treatment not been implemented with the number of reported crashes in the after period with the treatment. The EB method uses the safety performance of sites similar to treated sites for estimating the expected number of crashes had the treatment not been implemented. Particularly, it combines the observed number of crashes at a site with the estimated number of crashes of similar sites. The estimation is usually obtained from the SPFs. The steps are described below.

Step 1: Estimate the Expected Number of Crashes in the Before Period. Using an SPF and a set of CMFs, the expected number of crashes for a site can be estimated using Equation 1.

$$E\left[\hat{k}_i\right] = t \times C \times f(ADT, length) \times p \times CMF_1 \times \cdots \times CMF_k \tag{1}$$

where:

$E\left[\hat{k}_i\right]$ = predicted number of target crashes (e.g., SVROR + OD) for site i;

t = duration of the study period (usually in number of years);

C = calibration factor;

$f(ADT, length)$ = SPF for a set of base conditions, which only includes the traffic volume (represented as ADT) and segment length. This is consistent with procedures in the current *Highway Safety Manual* (*HSM*) (*24*);

p = proportion of target crash type;

$CMF_1, \ldots, CMF_k$ = a set of CMFs for other variables.

The EB method estimates the expected number of crashes before implementation of the countermeasure at each treatment site and the variance of $E[\hat{k}_i|K_i]$. (Note that "^" refers to an estimate of a variable.) The estimate $E[\hat{k}_i|K_i]$ is calculated by combining the predicted crashes $(E[k]_i)$ with the observed count of crashes (K_i) in the before period, and is given as follows:

$$E[\hat{k}_i|K_i] = \hat{w}_i \cdot E[\hat{k}_i] + (1 - \hat{w}_i) \cdot K_i \tag{2}$$

where:

$E[\hat{k}_i|K_i]$ = EB estimate of the expected number of crashes for site i;

$\hat{w}_i$ = weight factor; and,

K_i = observed number of crashes

The variance of the estimate is given as

$$Var\left[E[\hat{k}_i|K_i]\right] = (1 - \hat{w}_i) \cdot E\left[\hat{k}_i|K_i\right] \tag{3}$$

Details of SPFs, CMFs, calibration factor, proportion of target crash type, and weight factor are documented in the next section.

Step 2: Calculate the Ratio of the After Period Crash Estimate to the Before Period Estimate. With the SPFs used in Step 1, estimate the expected number of crashes $(E[z_i])$ in the after period at each treatment site. The ratio of the after period crash estimate to the before period estimate (P_i) is calculated as:

$$P_i = \frac{E\left[\widehat{Z}_i\right]}{E\left[\widehat{k}_i\right]} \tag{4}$$

Step 3: Obtain the Estimated Crashes ($\hat{\pi}_i$) and Its Estimated Variance. Calculate the estimated crashes during the after period that would have occurred without implementing the countermeasure (i.e., profile pavement markings).

The estimated number of crashes ($\hat{\pi}_i$) is given by:

$$\hat{\pi}_i = P_i \times E[\hat{k}_i|K_i] \tag{5}$$

The estimated variance of $\hat{\pi}_i$ is given by:

$$Var[\hat{\pi}_i] = P_i \times (1 - \hat{w}_i) \times E\left[\hat{k}_i|K_i\right] \tag{6}$$

Step 4: Compute the Sum of the Estimated and Observed Crashes Over All Sites in the Treatment Group. The number of after period crashes for a group of sites had the treatment not been implemented at the treated sites is given as:

$$\hat{\pi} = \sum_{i=1}^{J} \hat{\pi}_i \tag{7}$$

where J represents the total number of sites in the treatment group, and $\hat{\pi}$ is the estimated after period crashes at all treated sites had there been no treatment, as described above.

Step 5: Compute the Sum of the Actual Crashes Over All Treated Sites. For a treated site, crashes in the after period are influenced by the implementation of the treatment. The safety effectiveness of a treatment is assessed by comparing the actual crashes with the treatment with the estimated crashes without the treatment. The actual number of after period crashes for a group of treated sites is given as:

$$\hat{\lambda} = \sum_{i=1}^{J} L_i \tag{8}$$

where L_i is the crash frequency during the after period at site i. The estimate of $\hat{\lambda}$ is equal to the sum of the observed number of crashes at all treated sites during the after study period.

Step 6: Estimate $Var\left[\hat{\lambda}\right]$ and $Var[\hat{\pi}]$. Based on the assumption of a Poisson distribution, the estimate of variance of $\hat{\lambda}$ is assumed to be equal to L. The estimate of variance of $\hat{\pi}$ can be calculated from the equations as follows:

$$Var\left[\hat{\lambda}_i\right] = L_i \tag{9}$$

$$Var\left[\hat{\lambda}\right] = \sum_{i=1}^{J} Var\left[\hat{\lambda}_i\right] \tag{10}$$

$$Var[\hat{\pi}_i] = P_i \cdot (1 - \hat{w}_i) \cdot E\left[\hat{k}_i|K_i\right] = (1 - \hat{w}_i) \cdot \hat{\pi}_i \tag{11}$$

$$Var[\hat{\pi}] = \sum_{i=1}^{j} Var[\hat{\pi}_i] \quad (12)$$

Step 7: Compute the Safety Effectiveness of the Treatment. The CMF is estimated as the ratio of safety with the treatment to what it would have been without the treatment.

$$\widehat{CMF} = \frac{\frac{\hat{\lambda}}{\hat{\pi}}}{1 + \frac{Var[\hat{\pi}]}{\hat{\pi}^2}} \quad (13)$$

The percentage change (known as crash reduction factor, CRF) in the number of target crashes due to the treatment is calculated by 100(1-$\widehat{CMF}$)%. If $\widehat{CMF}$ is less than 1, then the treatment has a positive safety effect. The estimated variance and standard error of the estimated safety effectiveness are given by:

$$Var\left(\widehat{CMF}\right) = \widehat{CMF}^2 \times \frac{\left(\frac{1}{\hat{\lambda}} + \frac{Var[\hat{\pi}]}{\hat{\pi}^2}\right)}{\left(1 + \frac{Var[\hat{\pi}]}{\hat{\pi}^2}\right)} \quad (14)$$

$$s.e.\left(\widehat{CMF}\right) = \sqrt{Var\left(\widehat{CMF}\right)} \quad (15)$$

The approximate 95% confidence interval for CMF is given by adding and subtracting $1.96 \times s.e.\left(\widehat{CMF}\right)$ from $\widehat{CMF}$. If the confidence interval contains the value 1, then no significant effect has been observed.

For the derivation of Equations 2–15, readers are referred to (*5*, *25*, *26*).

SPFs and CMFs

As discussed in the previous section, the EB estimate is essentially a combination of two sources: prediction from SPFs and CMFs, and observed number of crashes. The former depends on the SPFs and CMFs, and on the weight factor (i.e., w_i in Equation 2), which is usually derived from the SPFs. In most previous studies utilizing the EB approach, SPFs were specifically developed using control or reference sites. However, the number of control sites used to develop an accurate over-dispersion parameter of an SPF is usually problematic, especially when the sample mean is relatively low. For instance, Khan et al. (*17*) used just 53 reference sites for developing an SPF. When the sample size is low, the estimated dispersion parameter is highly biased (*27*), hence it will influence the EB estimates. Meanwhile, Persaud and Lyon (*28*) found that the final estimates heavily depended on the chosen reference group. When these two problems are combined, the CMF estimates from an EB approach may be highly biased. To overcome this problem, this study used two well-developed SPFs: (a) SPFs and CMFs for two-lane rural highways included in Chapter 10 of the *HSM* (HSM SPF) (*24*); and (b) SPFs and CMFs in the Texas Highway Safety Design Workbook (Texas SPF) (*29*). Note that the latter was selected because the treated sites in this study were all in Texas, as presented in the Data Description section.

HSM *SPF.* The SPFs for base conditions in the *HSM* for all crashes on two-lane roads is given as:

$$\mathrm{E[k]} = \mathrm{ADT} \times \mathrm{L} \times 365 \times 10^{-6} \times e^{-0.312} \quad (16)$$

where:

E[k] = the predicted number of all crashes per year for base conditions;

L = Segment length (mi); and

ADT = average daily traffic.

The weight $\hat{w}_i$ is given as

$$\hat{w} = \frac{1}{1 + \frac{E[\hat{k}] \times 0.236}{L}} \quad (17)$$

It is worth mentioning that the *HSM* requires the calibration factor to be adjusted for local conditions. This study followed the procedure documented in Part C of the *HSM* and estimated the calibration factor.

Texas SPF. The SPFs for base conditions in the Texas Highway Safety Design Workbook for fatal and injury (FI) crashes on two-lane roads can be given as:

$$\mathrm{E[k]} = 0.0537 \times (0.001ADT)^{1.30} \times \mathrm{L} \quad (18)$$

The weight $\hat{w}_i$ is given as

$$\hat{w} = \frac{1}{1 + \frac{E[\hat{k}]}{L \times 15.3}} \quad (19)$$

For both *HSM* SPF and Texas SPF, CMFs for a variety of variables (i.e., lane width, shoulder width, presence of passing lane, driveway density, and horizontal clearance) were used in the analyses. For specific values of these CMFs, readers are referred to Chapter 10 of *HSM* (*24*) and Chapter 3 of the Texas Highway Safety Design Workbook (*29*).

Data Description

This section documents the roadway and traffic data for treated sites, and the distribution of crash types and severity levels on rural two-lane highways in Texas.

Table 1. Summary Statistics of Treated Sites (33 Sites)

Variable	Min	Max	Mean	SD
Length (mi)	1.45	12.55	5.75	2.50
Shoulder width (ft)	0	10.0	4.94	3.10
Lane width (ft)	10.0	23.8	13.07	2.98
Median width (ft)	0	67	2.0	11.4
Horizontal clearance (ft)	22.5	30	28.0	1.71
Driveway density (# per mi)	4.71	250.0	22.8	45.70
Passing lane presence	0	1	0.06	0.12
ADT (Before)	237.0	12681.4	3879.7	3135.5
ADT (After)	213.3	12112.0	3736.3	2999.1
Number of SVROR + OD crashes (Before)	1	30	9.4	6.1
Number of SVROR + OD crashes (After[a])	0	16	4.7	3.4
Crash rate of SVROR + OD (Before)[b]	3.02	7.04	13.04	17.24
Crash rate of SVROR + OD (After)[b]	0.0	226.21	10.28	38.94

[a]The after period for each site varies.
[b]Crash rate is defined as number of crashes per million vehicle miles of travel.

Table 2. Distribution of Crash Types and Severity Levels on Rural Two-Lane Highways in Texas

	All types of crashes		SVROR + OD crashes	
Severity	Frequency[a]	Percentage	Frequency[a]	Percentage
K	2,774	3.0	2,321	3.9
A	6,527	7.0	5,172	8.8
B	14,411	15.4	11,049	18.7
C	11,504	12.3	8,004	13.6
PDO	58,112	62.3	32,481	55.0
Total	93,328	100	59,027	100

[a]Frequency of crashes from 2012 to 2014 (intersection and driveway related crashes have been excluded).

Roadway and Traffic Data at Treated Sites

The profile pavement marking and rumble bar installation locations were provided by the Texas Department of Transportation (TxDOT), Atlanta District. In total there are 33 sites that have been treated with audible lane departure warning treatments on 189 miles of rural two-lane highways in the Atlanta District. The treatment implementation dates were matched through TxDOT's project daily work report. A three-year period prior to the project installation beginning date is used as the before period for each treated site. The after period considered is from the day the project ended to three years later or October 31, 2016 (the latest date of crash data available to the research team), whichever is earlier.

The study collected crash data from the Crash Records Information System (CRIS) maintained by TxDOT in the before and after periods for each site. In addition to total target crashes, crash frequency by different severities was collected. This is important because it helps in understanding what severities are mostly influenced. Also, the Texas SPF estimates FI crashes only. The following five crash severity levels were considered: fatal (K), incapacitating injury (A), non-incapacitating injury (B), minor injury (C), and property damage only (PDO).

The TxDOT's Road-Highway Inventory Network (RHINO) database was used to extract variables such as traffic volume (ADT), surface width, shoulder width, number of lanes, and functional classification. Some specific roadway characteristics that are not included in the RHINO database were identified using Google Earth, including roadside clearance, presence of passing lane, and density of driveways. The summary statistics of the dataset are shown in Table 1.

Proportion of Target Crash Type

The proportion of crash types and severity levels on two-lane rural highways in Texas from 2012 to 2014 were calculated, as shown in Table 2.

Based on the distribution, the proportion of target crash type (i.e., p in Equation 1) can be calculated. For example, if the target crash type is SVROR + OD (PDO) crashes, they account for 34.8% of all crashes (32,481/93,328 = 34.8%).

Table 3. EB Analysis Results using *HSM* SPF

Site ID	All severity			FI			PDO		
	Obs.	Est.	SD	Obs.	Exp.	SD	Obs.	Est.	SD
1	1	2.41	0.52	1	1.19	0.25	0	1.30	0.29
2	16	20.66	2.67	4	8.00	1.19	12	9.53	1.42
3	4	5.49	1.24	0	2.65	0.65	4	3.09	0.76
4	3	8.00	1.41	1	3.51	0.70	2	4.27	0.84
5	5	6.30	1.34	1	2.97	0.70	4	3.46	0.81
6	6	2.82	0.49	3	1.26	0.23	3	1.53	0.28
7	6	7.34	0.87	2	3.46	0.44	4	4.08	0.52
8	5	8.11	1.44	2	3.17	0.69	3	4.53	0.89
9	9	9.24	1.98	5	5.08	1.14	4	5.26	1.24
10	1	4.00	1.40	1	2.20	0.82	0	2.51	0.94
11	3	7.95	1.20	2	4.09	0.66	1	4.50	0.74
12	10	5.50	0.96	3	2.82	0.52	7	2.73	0.55
13	6	6.29	1.11	1	3.10	0.57	5	3.28	0.64
14	3	6.03	1.11	1	2.51	0.53	2	3.48	0.68
15	2	3.76	0.92	2	1.64	0.43	0	1.70	0.48
16	12	16.22	2.46	4	8.15	1.33	8	9.21	1.52
17	4	6.38	1.17	2	3.41	0.65	2	3.55	0.71
18	9	4.48	0.80	5	2.00	0.39	4	2.54	0.48
19	4	15.64	2.68	1	7.73	1.44	3	8.67	1.64
20	5	8.06	1.03	1	3.59	0.50	4	4.18	0.58
21	5	5.93	0.91	2	2.62	0.44	3	2.85	0.50
22	7	14.00	2.68	3	8.19	1.62	4	8.23	1.74
23	3	2.16	0.42	3	1.01	0.21	0	1.19	0.25
24	2	3.47	0.52	0	1.58	0.25	2	1.73	0.29
25	4	2.76	0.38	4	1.24	0.18	0	1.42	0.21
26	1	1.45	0.24	0	0.64	0.11	1	0.82	0.14
27	4	5.26	0.95	2	2.29	0.43	2	2.57	0.51
28	1	0.33	0.07	1	0.15	0.03	0	0.16	0.04
29	5	3.80	0.54	2	1.70	0.25	3	2.04	0.30
30	0	0.50	0.12	0	0.22	0.05	0	0.28	0.07
31	3	1.12	0.16	0	0.52	0.08	3	0.58	0.09
32	3	0.24	0.05	2	0.08	0.02	1	0.12	0.03
33	3	1.01	0.23	2	0.40	0.10	1	0.55	0.13
Total	155	196.7		63	93.1		92	106.0	

Note: The calibration factor was calculated as 0.802; Obs. = observed crash number; Est. = estimated crash number.

Results

The following sections describe the EB estimates of CMFs for profile pavement markings and rumble bars.

HSM *SPF*

The results of the EB analysis for all, FI, and PDO SVROR + OD crashes, respectively, on two-lane rural highways are shown in Table 3. The observed number of target crashes in the after period, estimated number of crashes using the EB method, as well as standard deviation of the EB estimates for each site, are listed in Table 3. As can be seen, the value of standard deviation is relatively small compared with the corresponding estimated crash number, indicating the results are stable.

Table 3 shows that there are in total 155 crashes reported in the after study period at these 33 treated sites. The analysis results indicate that if the treatments had not been installed, the estimated number of the target crashes would have been 196.7 in the after study period based on *HSM* SPF and CMFs. The expected FI and PDO SVROR + OD crashes in the after period are 93.1 and 106.0, respectively, had the treatments not been implemented, compared with the observed numbers in the after period which are 63 and 92, respectively. Overall, all three types of crashes were reduced after the implementation of alternative audible lane departure warning treatments. Further development of CMFs is discussed in the following section.

Texas SPF

The results of the EB analysis with Texas SPF and CMFs for FI SVROR + OD crashes on two-lane rural

Table 4. EB Analysis Results using Texas SPF

	FI		
Site ID	Obs.	Estimated	SD
1	1	1.34	0.16
2	4	5.02	0.45
3	0	4.69	0.65
4	1	5.55	0.64
5	1	5.10	0.69
6	3	0.82	0.08
7	2	4.78	0.34
8	2	4.55	0.59
9	5	8.55	1.09
10	1	4.65	0.97
11	2	5.92	0.54
12	3	3.59	0.39
13	1	2.62	0.28
14	1	2.89	0.34
15	2	1.28	0.20
16	4	12.32	1.14
17	2	4.11	0.44
18	5	2.27	0.24
19	1	9.86	1.05
20	1	3.67	0.29
21	2	2.77	0.27
22	3	1.11	0.12
23	3	0.86	0.10
24	0	1.11	0.10
25	4	0.73	0.06
26	0	0.39	0.04
27	2	2.19	0.23
28	1	0.07	0.01
29	2	1.23	0.10
30	0	0.11	0.01
31	0	0.30	0.02
32	2	0.03	0.00
33	2	0.23	0.03
Total	63	104.7	

Note: The calibration factor was calculated as 1.254.

highways are shown in Table 4. The observed number of crashes, expected with EB method, and standard deviation for each site are also provided. As can be seen, with the Texas SPF, the estimated FI SVROR + OD crashes at these 33 sites would be 104.7 had the treatments not been implemented, whereas the observed number is 63 in the after period. Similar to that using *HSM* SPF and CMFs, the FI SVROR + OD crashes decreased after installing profile pavement markings and rumble bars.

Using *HSM* SPF and CMFs, it is estimated that alternative audible lane departure warning treatments reduced SVROR + OD crashes (all severity) by 21.3%. The standard deviation is 7%, which makes the estimate statistically significant at 95% level. The estimated CMF for FI SVROR + OD crashes is 0.68, and the standard deviation is 0.09, meaning that the CRF is 32.5%. This result is also statistically significant at 95% level. For PDO crashes, the CRF is 13%. However, it is not significant at 95% level.

While using Texas SPF and its CMFs, the CMF for audible lane departure warning systems is estimated as 0.60 (CRF = 39.9%) with a standard error of 0.08. This result is statistically significant at the 95% level. This is close to the estimate obtained with *HSM* SPF (i.e., 0.68).

The results related to CMFs are summarized in Table 5. The EB analyses indicate that, after installing the alternative audible lane departure warning treatments on two-lane highways, all SVROR + OD crashes are expected to be reduced by about 21.3%, FI SVROR + OD crashes would be reduced by about 32.5% to 39.9%. Except for PDO crashes, all other results are statistically significant at 95% level. In short, installing the treatments is expected to reduce SVROR + OD crashes on two-lane highways.

Sensitivity Analysis of Calibration Factor

To eliminate the issue of low sample size of control sites, this study tried to utilize two well-established sets of SPFs rather than developing a new one. As suggested by

Table 5. Summary Results of the EB Analysis

	HSM SPF			Texas SPF
Variables	Total crashes	FI crashes	PDO crashes	FI crashes
Predicted crashes ($\hat{\pi}$)	196.7	93.1	106	104.7
	(53.5)	(15)	(19.5)	(7.7)
Estimated crashes ($\hat{\lambda}$)	155	63	92	63
	(12.4)	(7.9)	(9.6)	(7.9)
CMF	**0.79**	**0.68**	0.87	**0.60**
	(0.07)	(0.09)	(0.10)	(0.08)
CRF	**21.3**	**32.5**	13.3	**39.9**
	(7.0)	(8.9)	(9.7)	(7.7)

Note: Bold indicates statistically significant at the 95% level. Value in parenthesis is the standard error of the estimate.

the *HSM*, SPFs need to be calibrated in most cases, especially if one was developed using crash data in a different jurisdiction. This study followed the calibration procedure documented in the *HSM*, and the calibration factors were estimated as 0.802, and 1.254, respectively, in the two sets of SPFs (i.e., *HSM* SPF and Texas SPF). It is worth mentioning that both SPFs were calibrated with data in the before period at treatment sites, which might have over-estimated the calibration factors. A further analysis on the sensitivity of calibration factor on estimated CMFs was conducted.

The calibration factor was assumed to be some constant between 0.5 and 1.5, and the CMFs were then estimated using the same steps described in previous sections. The relationship between CMF and assumed calibration factor are illustrated in Figure 2. Figures 2*a*, *b*, and *c* show the results of CMFs for all SVROR + OD with *HSM* SPF, FI SVROR + OD with *HSM* SPF, and SVROR + OD with Texas SPF, respectively. As expected, the calibration factor influences CMF estimates. It can be seen that, for all three scenarios, the estimated CMF decreases as the calibration factor increases. Meanwhile, the standard deviation of estimated CMF also decrease, indicating that the quality of CMF is improved.

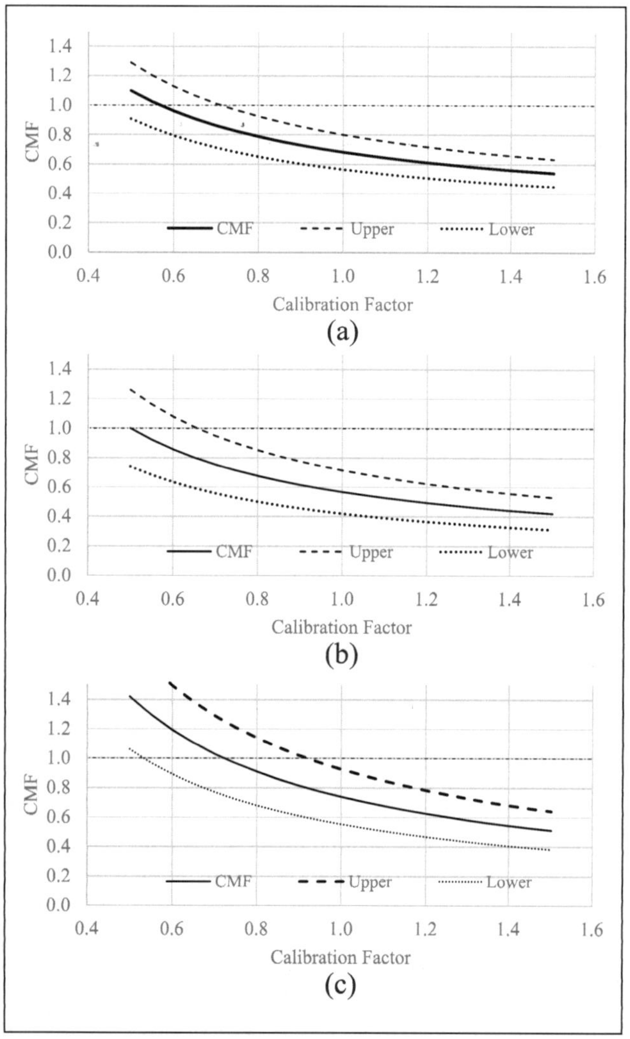

Figure 2. Sensitivity analysis of calibration factor on estimated CMFs : (*a*) All SVROR + OD with *HSM* SPF; (*b*) FI SVROR + OD with *HSM* SPF; (*c*) FI SVROR + OD with Texas SPF.

For all SVROR + OD with *HSM* SPF, the estimated CMF will be smaller than 1.0 as long as the calibration factor is greater than about 0.55, and the result will be statistically significant at 95% level if the calibration factor is greater than about 0.70. For FI SVROR + OD with *HSM* SPF, CMF remains under 1.0 when calibration factor is greater than 0.5, and statistically significant at 95% level if the calibration factor is greater than about 0.65. For FI SVROR + OD with Texas SPF, the estimated CMF will be smaller than 1.0 when the calibration factor is greater than about 0.70, and it is statistically significant at 95% level if the calibration factor is greater than about 0.90.

Conclusions and Discussions

This study has examined the effectiveness of profile pavement markings and rumble bars in reducing the number of SVROR and OD crashes on rural two-lane highways in Texas using the EB before–after analysis method. The results of this study demonstrate the safety benefits of installing these treatments in reducing lane departure crashes. These alternative audible lane departure treatments, implemented on about 200 miles of rural two-lane highways in Atlanta District, Texas, were used as the primary data source for the analysis. Three years of crash data before the implementation and crash data in the after periods were extracted from TxDOT CRIS. Roadway characteristics and traffic volume data were collected from TxDOT RHINO and Google Earth. The study found a 21.3% reduction in SVROR and OD (all severity) crashes, and 32.5% to 39.9% reduction in SVROR and OD (FI) crashes after installing the treatments. Both results are statistically significant at the 95% level. This finding generally agrees with previous studies (e.g., (*22*, *30*, *31*)). Profile pavement markings and rumble bars showed similar safety effects as rumble strips. For example, Torbic et al. (*12*) reported a CMF of 0.53 for edgeline rumble strips on rural two-lane undivided highways with a shoulder less than 5 feet. PDO SVROR and OD crashes also showed the trend of fewer crashes, but the result is not statistically significant. Considering the relatively low cost of installing these treatments, the results obtained in this study suggest the use of profile pavement markings and rumble bars on rural two-lane highway segments where SVROR and

OD collisions are a safety concern, especially where milled or rolled rumble strips are not applicable.

Compared with other studies that have utilized the EB method for estimating treatment effects, this study adopted two well-established SPFs and CMFs (i.e., *HSM* and Texas Highway Safety Design Workbook SPFs) rather than developing an SPF based on comparison sites. This eliminated the low sample size bias in developing SPFs found in many previous studies (*17*, *28*). Safety analysts are recommended to consider adopting SPFs when conducting EB analysis with limited number of reference sites.

It is worth mentioning that, when estimating target crashes (i.e., by collision type and severity level), this study followed the procedure documented in the *HSM* which recommends using a fixed proportion for each collision type and severity level. Future work should consider estimating the crashes using target crash models.

Acknowledgments

The authors thank TxDOT for providing the data used in this paper and funding this study.

Author Contributions

The authors confirm contribution to the paper as follows: study conception and design: Adam M. Pike and Srinivas R. Geedipally; data collection: Lingtao Wu; analysis and interpretation of results: Lingtao Wu, Srinivas R. Geedipally, and Adam M. Pike; draft manuscript preparation: Lingtao Wu. All authors reviewed the results and approved the final version of the manuscript.

References

1. FHWA. *Strategic Approach to Roadway Departure Safety*. https://safety.fhwa.dot.gov/roadway_dept/strat_approach/. Accessed July 3, 2017.
2. Dissanayake, S., and U. Roy. Crash Severity Analysis of Single Vehicle Run-Off-Road Crashes. *Journal of Transportation Technologies*, Vol. 4, No. 1, 2014, p. 1.
3. Liu, C., and ,R. Subramanian. *Factors Related to Fatal Single-Vehicle Run-Off-Road Crashes*. Report DOT HS 811 232. URC Enterprises, Inc., Olney, Md., 2009.
4. Pike, A. M., B. T. Wilson, E. S. Park, S. R. Geedipally, and L. Wu. *Evaluation of Audible Lane Departure Warning Treatments for Seal Coat Road Surfaces*. Report FHWA/TX-17/0-6888-1. Texas A&M Transportation Institute, College Station, Texas, 2017.
5. Gross, F., B. Persaud, and C. Lyon. *A Guide to Developing Quality Crash Modification Factors*. Publication FHWA-SA-10-032. FHWA, U.S. Department of Transportation, Washington, D.C., 2010.
6. Wood, J. S., J. P. Gooch, and E. T. Donnell. Estimating the Safety Effects of Lane Widths on Urban Streets in Nebraska Using the Propensity Scores-Potential Outcomes Framework. *Accident Analysis & Prevention*, Vol. 82, 2015, pp. 180–191.
7. Gooch, J. P., V. V. Gayah, and E. T. Donnell. Quantifying the Safety Effects of Horizontal Curves on Two-Way, Two-Lane Rural Roads. *Accident Analysis & Prevention*, Vol. 92, 2016, pp. 71–81.
8. Wu, L., D. Lord, and Y. Zou. Validation of Crash Modification Factors Derived from Cross-Sectional Studies with Regression Models. *Transportation Research Record: Journal of the Transportation Research Board*, 2015. 2514: 88–96.
9. Hauer, E. Should Stop Yield? Matters of Method in Safety Research. Institute of Transportation Engineers Journal, Vol. 61, No. 9, 1991, pp. 25–31.
10. Shen, J., and A. Gan. Development of Crash Reduction Factors: Methods, Problems, and Research Needs. *Transportation Research Record: Journal of the Transportation Research Board*, 2003. 1840: 50–56.
11. Wu, L., Y. Zou, and D. Lord. Comparison of Sichel and Negative Binomial Models in Hot Spot Identification. *Transportation Research Record: Journal of the Transportation Research Board*, 2014. 2460: 107–116.
12. Torbic, D. J., J. M. Hutton, C. D. Bokenkroger, K. M. Bauer, D. W. Harwood, D. K. Gilmore, J. M. Dunn, J. J. Ronchetto, E. T. Donnell, H. J. Sommer, P. M. Garvey, B. Persaud, and C. Lyon. *NCHRP Report 641: Guidance for the Design and Application of Shoulder and Centerline Rumble Strips*. Transportation Research Board of the National Academies, Washington, D.C., 2009.
13. Patel, R. B., F. M. Council, and M. S. Griffith. Estimating Safety Benefits of Shoulder Rumble Strips on Two-Lane Rural Highways in Minnesota: Empirical Bayes Observational Before-and-After Study. *Transportation Research Record: Journal of the Transportation Research Board*, 2007. 2019: 205–211.
14. Sayed, T., P. deLeur, and J. Pump. Impact of Rumble Strips on Collision Reduction on Highways in British Columbia, Canada Comprehensive Before-and-After Safety Study. *Transportation Research Record: Journal of the Transportation Research Board*, 2010. 2148: 9–15.
15. Park, J., and M. Abdel-Aty. Development of Adjustment Functions To Assess Combined Safety Effects of Multiple Treatments on Rural Two-Lane Roadways. *Accident Analysis & Prevention*, Vol. 75, 2015, pp. 310–319.
16. Rys, M., D. Karkle, and E. Russell. *Study of KDOT Policy on Lane and Shoulder Minimum Width for Application of Centerline Rumble Strips*. Report No. K-TRAN:KSU-10-7. Kansas Department of Transportation, Topeka, 2012.
17. Khan, M., A. Abdel-Rahim, and C. J. Williams. Potential Crash Reduction Benefits of Shoulder Rumble Strips in Two-Lane Rural Highways. *Accident Analysis and Prevention*, Vol. 75, 2015, pp. 35–42.
18. Fu, H., and C. G. Wilmot. *Assessing Performance of Alternative Pavement Marking Materials*. Report 08-4TA. Louisiana Transportation Research Center, Baton Rouge, 2008.
19. Cairney, P. The Effect of Profile Line-Marking on In-Vehicle Noise Levels. In *Fatigue and Driving: Driver Impairment, Driver Fatigue and Driving Simulation* (L. Hartley, ed.), Taylor & Francis, London, 1995, pp. 127–132.

20. Lindly, J. K., and R. K. Wijesundera. *Evaluation of Profiled Pavement Markings*. Report FHWA/CA/OR. University Transportation Center for Alabama, 2003.
21. Kiesel, K. *Management of Audio Tactile Edge Line Markings*. Report No. 952. Australian Road Research Board, Australia, 2007.
22. Hatfield, J., S. Murphy, R. F. S. Job, and W. Du. The Effectiveness of Audio-Tactile Lane-Marking in Reducing Various Types of Crash: A Review of Evidence, Template for Evaluation, and Preliminary Findings from Australia. *Accident Analysis & Prevention*, Vol. 41, No. 3, 2009, pp. 365–379.
23. Montella, A. Safety Evaluation of Curve Delineation Improvements Empirical Bayes Observational Before-and-After Study. *Transportation Research Record: Journal of the Transportation Research Board*, 2009. 2103: 69–79.
24. AASHTO. *Highway Safety Manual*. American Association of State Highway and Transportation Officials, Washington, D.C., 2010.
25. Hauer, E. *Observational Before–after Studies in Road Safety: Estimating the Effect of Highway and Traffic Engineering Measures on Road Safety*. Pergamon, Tarrytown, New York, 1997.
26. Lord, D., and S. Geedipally. *Safety Effects of the Red-Light Camera Enforcement Program in Chicago, Illinois*. Lord Consulting, College Station, Texas., 2014.
27. Lord, D. Modeling Motor Vehicle Crashes Using Poisson-Gamma Models: Examining the Effects of Low Sample Mean Values and Small Sample Size on the Estimation of the Fixed Dispersion Parameter. *Accident Analysis & Prevention*, Vol. 38, No. 4, 2006, pp. 751–766.
28. Persaud, B., and C. Lyon. Empirical Bayes Before-After Safety Studies: Lessons Learned from Two Decades of Experience and Future Directions. *Accident Analysis & Prevention*, Vol. 39, No. 3, 2007, pp. 546–555.
29. Bonneson, J. A., and M. P. Pratt. *Roadway Safety Design Workbook*. Report FHWA/TX-09/0-4703-P2. Texas Transportation Institute, Texas A&M University System, College Station, Tex., 2009.
30. Park, J., and M. Abdel-Aty. Development of Adjustment Functions To Assess Combined Safety Effects of Multiple Treatments on Rural Two-Lane Roadways. *Accident Analysis & Prevention*, Vol. 75, 2015, pp. 310–319.
31. Job, R. F. S., and J. Hatfield. Review of the Evidence for Profile Line Markings as a Road Safety Countermeasure: II. What We Know and What We Need to Research. *Roadwise*, Vol. 15, 2004, pp. 9–14.

The Standing Committee on Traffic Control Devices (AHB50) peer-reviewed this paper (18-01098).

The opinions, findings and conclusions expressed in this paper do not necessarily reflect those of TxDOT.

Article

TRR
JOURNAL OF THE TRANSPORTATION RESEARCH BOARD

Transportation Research Record
2018, Vol. 2672(21) 41–50

DOI: 10.1177/0361198118794544
journals.sagepub.com/home/trr
SAGE

Closed-Course Study to Assess the Conspicuity of Wrong-Way Driving Countermeasures

Melisa D. Finley[1] and Jeffrey D. Miles[2]

Abstract

Alcohol has been found to be the primary contributing factor in many wrong-way crash studies. Thus, Texas A&M Transportation Institute researchers conducted a night-time closed-course study to assess the conspicuity of select signing and pavement marking wrong-way driving countermeasures from the perspective of alcohol-impaired drivers. Lowering the sign height, making the sign larger (i.e., oversized), adding red retroreflective material to the sign support, or adding flashing red light-emitting diodes around the border of the sign did not improve the ability of the alcohol-impaired drivers to locate WRONG WAY signs in a search task. However, in post-task assessments it was the opinion of the participants that the oversized sign, red retroreflective material, and flashing lights made it easier to find the WRONG WAY sign. The participants also thought that these three countermeasures caught their attention more than the standard size WRONG WAY sign mounted at 7 ft. Researchers did not find a significant difference in the recognition time between standard and modified wrong-way arrow marking designs. The modified arrow had a narrower head that provided equivalent recognition while potentially reducing maintenance costs. The participants also similarly assessed the ease at which they could find the two arrow designs among the other markings. Thus, it appears that the modified design performed as well as the current design. Researchers recommend implementing the modified wrong-way arrow design as existing wrong-way arrows are replaced.

Recently, many transportation agencies have been investigating the characteristics of wrong-way driving (WWD) crashes on freeways and the effectiveness of countermeasures intended to reduce wrong-way entries. In many states, alcohol was found to be the primary contributing factor in the majority of WWD crashes. The most common countermeasures include signs and pavement markings. Often methods for enhancing the conspicuity of standard signs are employed to attract a wrong-way driver's attention to the sign and its message. However, the impact of alcohol on a driver's ability to recognize traffic control devices and read/recognize the message was unknown.

Texas A&M Transportation (TTI) researchers investigated the behaviors of alcohol-impaired drivers and their interpretations of various traffic control devices in two night-time closed-course studies. Although data in response to a simulated environment cannot be directly compared with data collected on an actual road, closed-course study data can be used to compare the relative differences in performance between the various treatments evaluated. The objectives of the first study were to determine:

- Where alcohol-impaired drivers look in the forward driving scene.
- The impact of alcohol on sign color recognition.
- The impact of alcohol on sign legibility distance.

Details concerning the first study can be found in Finley et al. (*1*).

The objective of the second study was to assess the conspicuity of select WWD signing countermeasures and a modified wrong-way arrow design from the perspective of alcohol-impaired drivers. The scope of this paper is limited to the second study only. More information regarding all studies can be found in the project report (*2*).

Background

In 2012, the National Transportation Safety Board (NTSB) released a special investigation report on WWD

[1]Texas A&M Transportation Institute, College Station, TX
[2]Texas Department of Transportation, Bryan, TX

Corresponding Author:
Address correspondence to Melisa D. Finley: m-finley@tti.tamu.edu

on freeway and access ramps (*3*). The NTSB concluded that although wrong-way crashes occur relatively infrequently (accounting for only about 3% of all crashes on freeways), they are more likely to result in fatal and serious injuries than other types of highway crashes because the vast majority are head-on collisions. The NTSB also found the following:

- Entering an exit ramp in the wrong direction was the primary origin of wrong-way movements onto freeways.
- More than half of wrong-way drivers were impaired by alcohol.
- Wrong-way crashes occur more frequently at night and on the weekend.

Many other studies have identified similar trends (*4–13*).

A more recent query of 8 years of Fatality Analysis Reporting System crash data (2004–2011) (*14*) found that on average 58% of wrong-way crashes in the United States are related to driving under the influence of alcohol, drugs, or both substances. This was nearly twice the rate of alcohol or drug involvement for all fatal crashes. Furthermore, impaired male wrong-way drivers outnumbered impaired female wrong-way drivers by nearly 3:1.

Many studies have also identified a peak in wrong-way crashes around the typical closing time for establishments that serve alcohol (2:00 a.m.) (*3, 6, 8*). One Texas study (*9*) found that more than half of the wrong-way crashes occurred between midnight and 6:00 a.m. Studies in several other states have identified similar trends with respect to the early morning hours (*6, 10–12*).

In 2013, TTI researchers investigated the blood alcohol concentration (BAC) of 71 wrong-way drivers in Texas. Key findings regarding impairment were (*2*):

- Almost 90% had a BAC level equal to or greater than the legal limit (0.08 g/dL).
- Approximately 50% had a BAC level equal to or greater than twice the legal limit (0.16 g/dL).
- Approximately 10% had a BAC level equal to or greater than three times the legal limit (0.24 g/dL).
- The average BAC level was 0.18 g/dL (over twice the legal limit).

The 2009 *Manual on Uniform Traffic Control Devices* (*15*) contains information regarding the installation of DO NOT ENTER, WRONG WAY, ONE WAY, other regulatory signs, and pavement markings to deter wrong-way movements at exit ramps on freeways. At least one DO NOT ENTER and WRONG WAY sign must be placed at exit ramps facing a potential wrong-way driver. The standard height for both of these signs is 7 ft. However, these signs may be installed at a lower sign height to increase their visibility at night. Methods to increase the conspicuity of standard signs include, but are not limited to:

- Increasing the size of a sign.
- Placing a second identical sign on the opposite side of the roadway.
- Adding one or more red or orange flags above the sign.
- Adding a strip of retroreflective material to the sign support.
- Adding a beacon.
- Adding light-emitting diodes (LEDs) within the symbol or legend of the sign or border of the sign.

Wrong-way arrows on the pavement can also be used to notify a driver that they are traveling in the wrong direction. In Texas, wrong-way arrows are typically formed using retroreflective raised pavement markers (RRPMs). On several occasions, Texas Department of Transportation (TxDOT) personnel noted that the RRPMs comprising the wrong-way arrows have to be replaced frequently, especially those in the arrow head. Previous TTI research (*9*) also confirmed that wrong-way arrows in Texas sometimes had RRPMs missing. Researchers suspected that when RRPMs are missing from the arrow head, drivers (especially those that are impaired) find it more difficult to discern the arrow shape.

Experimental Design

The following sections describe the study treatments, vehicles and instrumentation, study procedure, and participant demographics. This study was considered human subjects research by the Texas A&M University Office of Human Subject Protection. For this reason, all participant recruiting materials and questions, as well as the study protocol, were reviewed and approved by the Texas A&M Institutional Review Board (IRB).

Treatments. The primary focus of the study was WRONG WAY signs. The WRONG WAY sign treatments included:

- Standard size signs.
- Oversized signs.
- Standard size signs with red retroreflective material on the sign support.
- Standard size signs with flashing red LEDs around the border of the sign.

Figure 1. Example of a sign treatment array.

At the time of this study, TxDOT was not lowering the height of WRONG WAY signs. However, the North Texas Tollway Authority was experimenting with WRONG WAY and DO NOT ENTER signs mounted at 2 ft. Therefore, all of the signing treatments were evaluated at two heights (7 ft and 2 ft).

To reduce learning effects, researchers used three lateral sign positions (left, middle, and right) and included distractor signs (i.e., YIELD, STOP, or DO NOT ENTER) in the sign treatment arrays. In some arrays, a distractor sign, instead of the WRONG WAY sign, included the conspicuity element (e.g., red retroreflective material on the sign support). Figure 1 contains an example of a sign treatment array that includes a standard size WRONG WAY sign mounted at 7 ft with red retroreflective material on the sign support.

The study also investigated two designs of wrong-way arrows constructed of RRPMs:

- The standard TxDOT design where the RRPMs at the end of the head are 24 in. from the stem (Figure 2*a*).
- A modified design where the RRPMs at the end of the head are located 18 in. from the stem in the hopes of reducing maintenance while preserving the shape of an arrow (Figure 2*b*).

To reduce learning effects, researchers again used three lateral positions (left, middle, and right) and included two distractor markings (i.e., a Y and a T) in the pavement marking treatment arrays. Figure 3 contains examples of the standard TxDOT arrow and the modified arrow in conjunction with the two distractor markings.

Overall, researchers created four treatment orders that included the sign and pavement marking treatments. All sign treatments were always seen before the arrow treatments. To reduce learning effects, each participant saw a different treatment order at each BAC level (i.e., 0.00, 0.04, 0.08, and 0.12 g/dL). In addition, researchers tried to balance the treatment orders across the four BAC levels. Additional details on the design of the treatments can be found in the final report (*2*).

Vehicles and Instrumentation. The study used one instrumented state-owned vehicle (2005 Dodge Grand Caravan) with the headlights aimed according to the manufacturer's instructions. Figure 4 shows the in-vehicle equipment used. Participants wore occlusion glasses that blocked their vision until the start of the study and in between treatment arrays (Figure 5). The study administrator controlled the occlusion glasses via a laptop computer, whereas the participant controlled the glasses with a handheld unit.

Study Procedure. Researchers recruited potential participants from the Bryan/College Station, Texas, area. All participants were required to have a valid driver's license with no night-time restrictions. In addition, the participants could not be color blind and had to meet the minimum levels of acceptable vision (20/40). Only male participants were used since the study involved consuming alcohol and the research team wanted to avoid potential medical complications associated with alcohol and pregnancy. Participants had to be at least 21-years old (legal drinking age in Texas). Researchers set a maximum weight limit in an effort to minimize the time needed for alcohol consumption and to maintain a similar alcohol consumption period for all participants.

Researchers conducted the study in three parts: pre-screening, Part 1, and Part 2. Each part was conducted on a different date. The pre-screening took about an hour and was conducted during the day. Upon arrival, a researcher verified that each participant met the study criteria. Each participant also completed two assessments that provided information about his typical alcohol consumption and behavior. Researchers used the answers to these assessments to identify naïve drinkers, individuals that may be at risk of alcohol dependence, and individuals that were alcohol dependent. Individuals in these three categories were not allowed to participate. Participants that met all the pre-screening criteria and

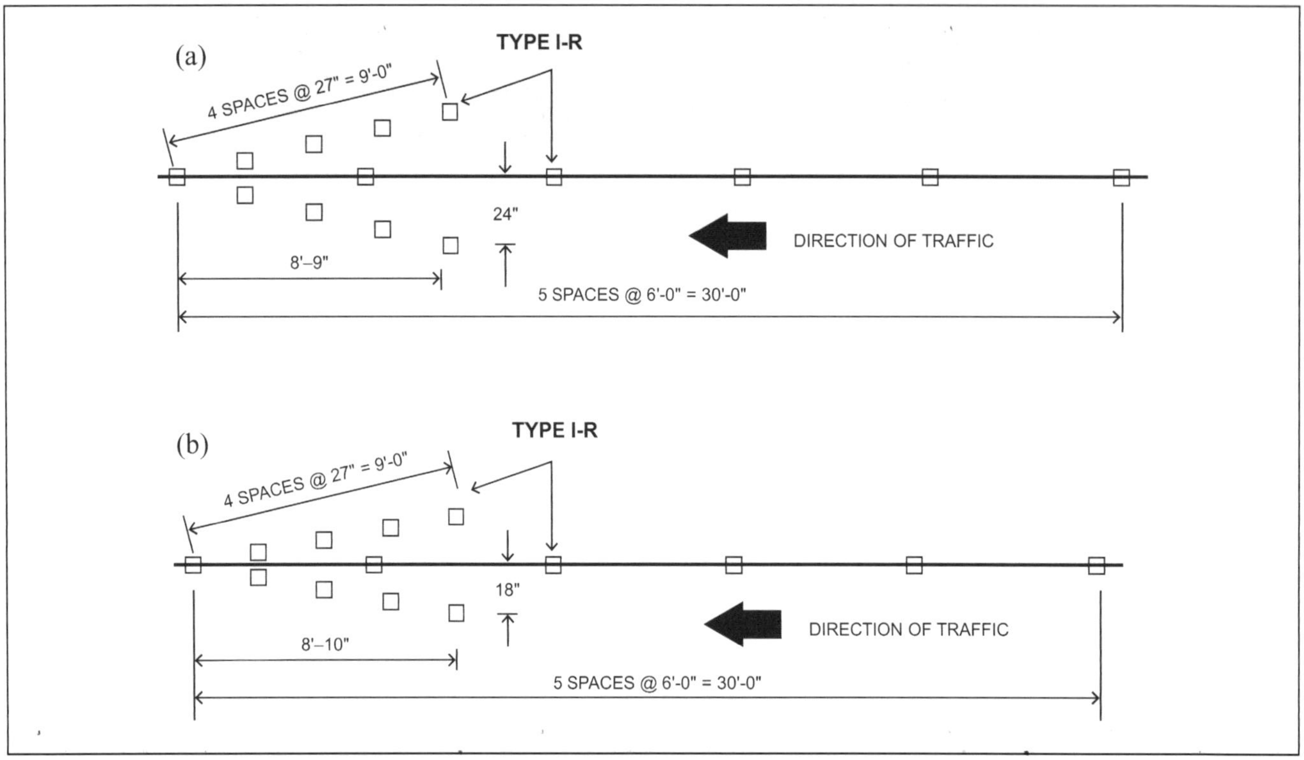

Figure 2. Wrong-way arrow designs studied: (*a*) TxDOT wrong-way arrow typical standard (*16*); and (*b*) modified wrong-way arrow.

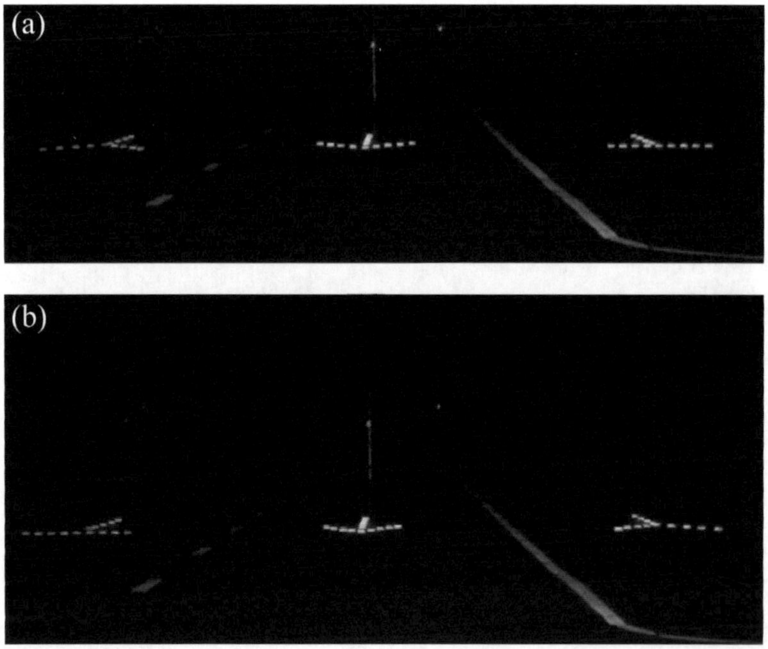

Figure 3. Examples of pavement marking treatment array: (*a*) TxDOT design (from left to right: Y, standard arrow, T); and (*b*) modified design (from left to right: T, modified arrow, Y).

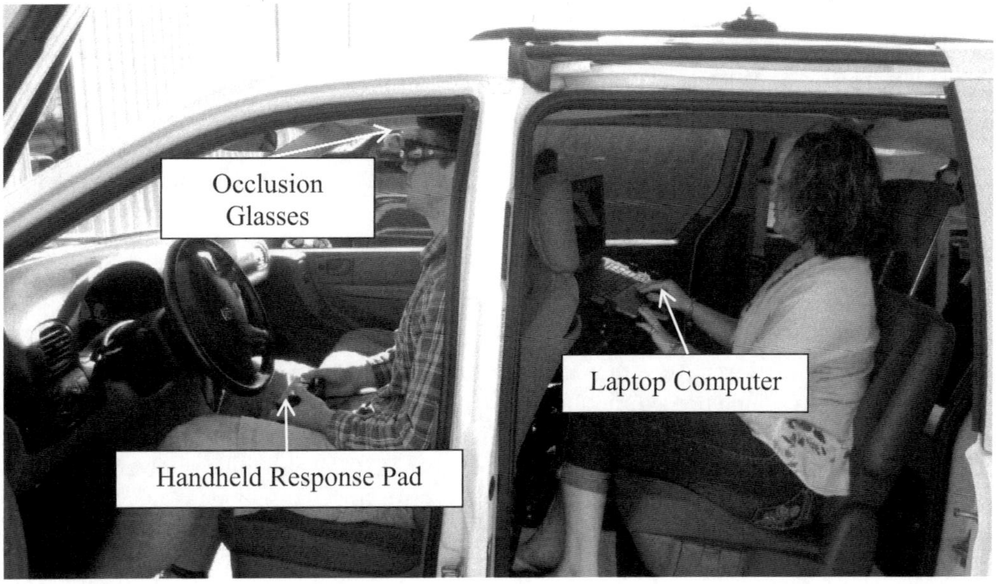

Figure 4. In-vehicle equipment.

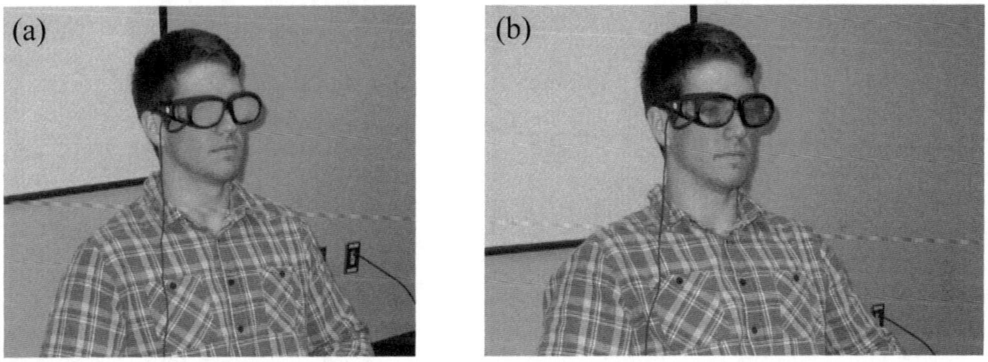

Figure 5. Occlusion glasses: (*a*) glasses activated (blocking vision); and (*b*) glasses deactivated (not blocking vision).

agreed to complete the remaining parts of the study were scheduled for Part 1.

Part 1 was conducted at night at the Texas A&M University RELLIS Campus and took about 2 hours. Participants did not consume any alcohol. Before beginning the study, each participant had his BAC level determined by standard Department of Public Safety (DPS) breath sample equipment to ensure that his BAC level was zero. In addition, each participant interacted with a short demonstration of the occlusion glasses and response pad.

A study administrator then drove the participant to the study location on the closed course. Once parked at the study location, the participant moved into the driver's seat and put on the occlusion glasses. The vehicle remained stationary during the study and was positioned 240 ft away from the treatment arrays. Once the study administrator gave the participant the in-vehicle instructions and handheld response pad, the study administrator activated the occlusion glasses (i.e., blocked the participant's vision) via software on a laptop computer. When the first sign treatment array was ready, the study administrator deactivated (or cleared) the occlusion glasses, restoring the participant's field-of-vision. The participant then found the WRONG WAY sign in the treatment array and pushed the correct button (left, right, or middle) on the response pad. When a button was pushed, the occlusion glasses were immediately activated, blocking the participant's field-of-vision. The time it took the participant to locate the WRONG WAY sign and respond was recorded in the software program. The study administrator then asked the participant whether the task of locating the WRONG WAY sign among the other treatments was easy or difficult and why. The study administrator marked the participant's response on a standard form. This procedure was repeated multiple times until the participant had viewed all of the sign treatment arrays.

Next, the participant located arrow pavement markings in treatment arrays using the same procedure. Once this was completed, the study administrator drove the participant back to the check-in building. Upon completion of Part 1 of the study, researchers scheduled the same participants for Part 2.

Part 2 was also conducted at night at the Texas A&M University RELLIS Campus and took approximately 10 hours. Participants were required to consume alcoholic beverages. Before consuming any alcohol, each participant had his BAC level determined by standard DPS breath sample equipment. Participants then consumed alcoholic beverages over approximately a 2-hour period until they reached a BAC level of 0.12 g/dL. Standard DPS breath sample equipment was used to monitor each participant's BAC level. At a minimum, each participant's BAC level was measured immediately prior to and after each closed-course task.

When each participant reached a BAC level of 0.12 g/dL, a study administrator drove the participant to the study location on the closed course. The state-owned vehicle, equipment, and participant tasks were the same as in Part 1. Once this was completed, the study administrator drove the participant back to the check-in building where the participant took a survey that assessed his opinion regarding the WRONG WAY sign treatments. The participants were then brought to a conference room where they had a comfortable place to sit, food, and nonalcoholic drinks. At a BAC level of 0.08 g/dL and 0.04 g/dL, each participant repeated the closed-course task. All participants had to remain on-site until their BAC level was less than 0.04 g/dL, at which time researchers drove them home.

The researchers acknowledge that this design was not counterbalanced by BAC level and thus learning and fatigue effects may have affected the results. The Texas A&M University IRB required that all participants undergo the part of the study without any impairment first. This was to ensure that participants could make a more informed decision about participating in the part of the study with impairment. In addition, due to time and budget constraints, the researchers could not have the same participants come for multiple sessions at the same time of day with different dosing levels. Since the study was conducted at night, researchers chose to use a descending BAC level in order for participants to begin consuming alcohol in the afternoon. This allowed the researchers to conduct the 0.12 g/dl run as soon as it was dark. Completing the closed-course tasks in descending BAC level also represents the experience of many intoxicated drivers.

Although data from Texas showed that approximately 77% of the alcohol-impaired drivers involved in WWD crashes had a BAC level equal to or greater than 0.12 g/dL, researchers did not study higher BAC levels because they wanted to minimize the medical risk to participants. In addition, Part 2 of the study typically took 10 hours per participant (ending between 3:00 and 4:00 a.m.). Higher BAC levels would have extended the study time.

Participants. Researchers prescreened 59 male participants. Fifteen participants did not meet the pre-screening criteria and thus were not allowed to participate. In addition, after completing the pre-screening, two participants decided not to participate, and eight were never scheduled. Of the remaining 34 participants, 34 completed Part 1 and 30 completed Part 2 of the study. After Part 1, researchers dropped one participant due to medical concerns, and three other participants were not available on the scheduled Part 2 study dates. The average age of the 30 participants was 25 and ranged from 21 to 48. The average visual acuity was 20/18, and the average weight was 194 lb.

Data Reduction and Analysis

Following data collection, the research team screened and reduced each participant's raw data into a fully formatted dataset to obtain the necessary information for analysis. During the data-screening process, researchers eliminated anomalous data (e.g., misidentifications and malfunctioning treatment/equipment).

Next, researchers reviewed the BAC-level data measured immediately before and after each closed-course task. Researchers computed the average BAC level of each participant for each closed-course task by averaging the two before and two after BAC-level measurements. A review of these data identified two outliers (one at 0.08 g/dL and one at 0.04 g/dL), which were removed from the final dataset. Table 1 contains the overall BAC level descriptive statistics.

Researchers calculated recognition times by subtracting the participants' identification time (i.e., the moment they pressed a button on the response pad) from the time the occlusion glasses restored the participants' field-of-vision (i.e., cleared by the study administrator). Researchers also reduced and analyzed the opinion data

Table 1. Overall BAC Level Descriptive Statistics

Target level (g/dL)	Average (g/dL)	Standard deviation (g/dL)	Minimum (g/dL)	Maximum (g/dL)
0.000	0.000	0.000	0.000	0.000
0.120	0.112	0.003	0.107	0.116
0.080	0.076	0.006	0.069	0.096
0.040	0.037	0.004	0.028	0.044

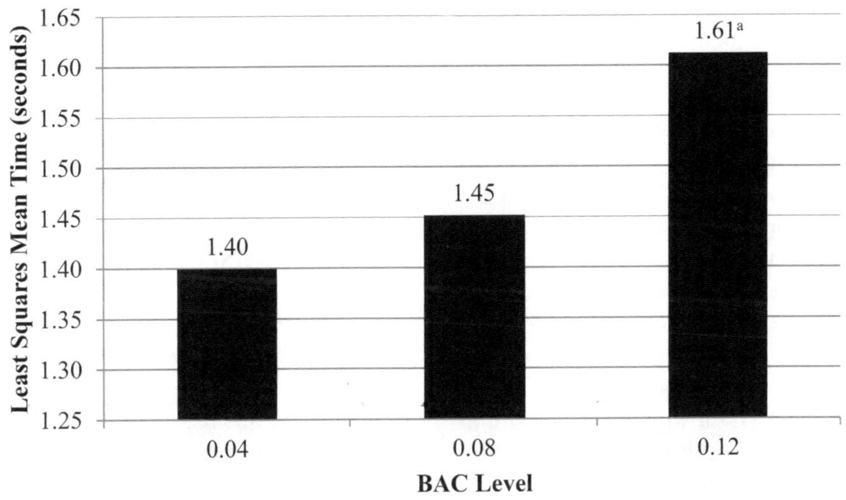

[a] BAC level 0.12 g/dL was statistically different from BAC levels 0.04 and 0.08 g/dL.

Figure 6. WRONG WAY sign recognition time by BAC level.

collected during each closed-course portion of the study and the BAC 0.12 g/dL survey.

Based on the initial exploratory analyses, researchers decided not to include BAC level 0.00 g/dL data and distractor sign/symbol data in the final models. Additional details regarding the initial exploratory analysis findings can be found in the final report (*2*).

When appropriate, researchers used the predicted values (least squares means) for each response variable to compare different treatments. When there are multiple factors in the model, it is not fair to make comparisons between raw cell means in data because raw cell means do not compensate for other factors in the model. The least squares means are the predicted values of the response variable for each level of a factor that have been adjusted for the other factors in the model. A 5% significance level was used for all statistical analyses.

Results

The following subsections contain the results of the analyses on the WRONG WAY sign treatments and two wrong-way arrow designs.

WRONG WAY signs. The final dataset included 1,108 observations, and the final model included the following main effects and two-way interactions:

- BAC level (0.12, 0.08, and 0.04 g/dL).
- Sign height (7 ft and 2 ft).
- Treatment (standard size, oversized, red retroreflective material on sign support, and flashing red LED border).
- BAC level * sign height.
- BAC level * treatment.
- Sign height * treatment.

This statistical analysis showed that only the BAC level was statistically significant ($p = 0.001$). Researchers used Tukey's Honestly Significant Difference (HSD) procedure to determine which BAC levels were statistically different. Figure 6 presents the predicted means for the WRONG WAY sign recognition time by BAC level. As expected, the recognition time increases as the BAC level increases. BAC level 0.04 g/dL was found to have the shortest predicted mean recognition time (1.4 s). However, this time was not significantly different from

Table 2. WRONG WAY Sign Participant Difficulty Assessment

Treatment	The task of locating the WRONG WAY sign among the other signs was difficult
Normal size sign	31%[a]
Oversized sign	17%
Normal size sign with red retroreflective sheeting on sign support	13%
Normal size sign with flashing red LEDs around border	13%

[a]Statistically different from the other percentages based on a test of proportions using an overall significance level of 5%.

the predicted mean recognition time for BAC level 0.08 g/dL (1.45 s). The longest predicted mean recognition time (1.61 s) occurred at BAC level 0.12 g/dL. This time was found to be significantly longer than those for BAC levels 0.08 g/dL and 0.04 g/dL. Thus, a BAC level of 0.12 g/dL did appear to negatively affect a motorist's ability to recognize the WRONG WAY signs among the other distractor signs independent of treatment.

Neither height nor treatment were found to be a significant variable in the model. Thus, the quantitative data yielded no significant differences between the conspicuity of a standard WRONG WAY sign mounted at 7 ft and the other treatments (i.e., lowered sign height, oversized sign, red retroreflective material on support, and flashing red LED border). Researchers believe the lack of background clutter and roadway lighting may have contributed to the similar response times among treatments.

In contrast, the qualitative difficulty assessment and participant opinion data did reveal some differences among the treatments. Table 2 shows the percentage of participants that thought the task of locating the WRONG WAY sign among the other signs was difficult by treatment. Participants thought the standard size WRONG WAY signs without any conspicuity element were more difficult to find than the oversized WRONG WAY signs, WRONG WAY signs with red retroreflective material on support, and WRONG WAY signs with flashing red LED border. Based on a test of proportions using a 5% significance level, the researchers determined that the difficulty percentage for the standard size WRONG WAY sign without any conspicuity element was statistically different from the other three percentages. No statistical differences were found between the other three treatments. In addition, data from the post-task survey administered to each participant at a BAC level of 0.12 g/dL revealed that between 84% and 92% of participants thought the oversized WRONG WAY signs, retroreflective material on the sign support, and flashing red LEDs caught their attention more than the standard sign (all treatments mounted at 7 ft).

Wrong-Way Arrow Pavement Markings. The final dataset included 135 observations, and the final model included the following main effects and their two-way interaction effect:

- BAC level (0.12, 0.08, and 0.04 g/dL).
- Treatment (standard arrow and modified arrow).

This statistical analysis showed that only the BAC level was statistically significant ($p = 0.016$). Researchers used Tukey's HSD procedure to determine which BAC levels were statistically different. Figure 7 presents the predicted means for the wrong-way arrow recognition time by BAC level. Again, as expected, the recognition time increases as the BAC level increases. The longest predicted mean recognition time (1.85 s) occurred at BAC level 0.12 g/dL. This time was found to be significantly more than the BAC level 0.04 g/dL time, but not the BAC level 0.08 g/dL time. Thus, a BAC level of 0.12 g/dL did appear to negatively affect a motorist's ability to recognize both types of wrong-way arrow pavement markings among the other distractor markings independent of treatment.

Treatment was not found to be a significant variable in the model. Thus, there was no statistical difference in the recognition time between the two wrong-way arrow marking designs. This means that the modified design performed as well as the current design. In addition, participants similarly assessed the ease at which they could find the two arrow designs among the other markings.

Conclusion

As alcohol has been found to be the primary contributing factor in many wrong-way crash studies, TTI researchers conducted a night-time closed-course study to assess the conspicuity of select WWD signing countermeasures and a modified wrong-way arrow pavement marking design using alcohol-impaired drivers.

Lowering the sign height, making the sign larger (i.e., oversized), adding red retroreflective material to the sign support, or adding flashing red LEDs around the border of the sign did not improve the ability of the alcohol-impaired drivers to locate WRONG WAY signs in a search task. However, in post-task assessments it was the opinion of the participants that the oversized sign, red

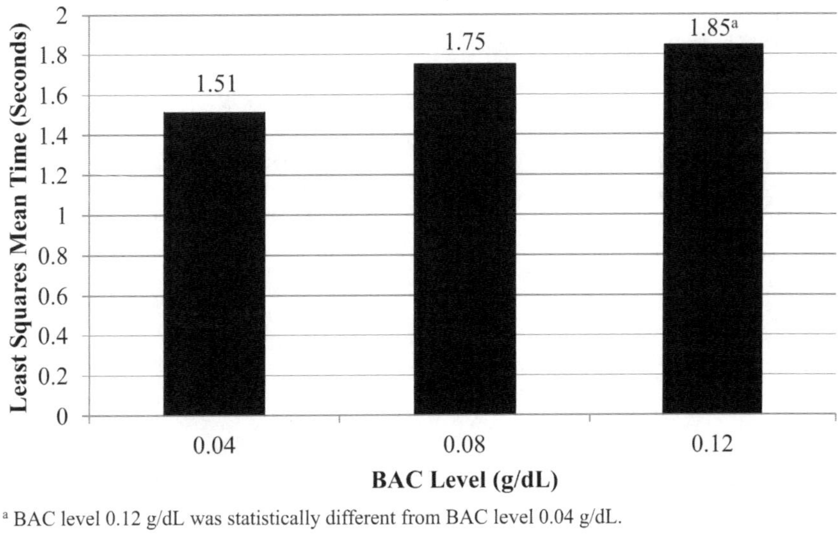

Figure 7. Wrong-way arrow recognition time by BAC level.

retroreflective material, and flashing lights did make it easier to find the WRONG WAY sign. The participants also thought that these three countermeasures caught their attention more than the standard size WRONG WAY sign mounted at 7 ft. Future research regarding the conspicuity of WWD signing and marking countermeasures should consider using background clutter and roadway lighting to better simulate a real-roadway application.

Researchers did not find a significant difference in the recognition time between the two wrong-way arrow marking designs. In addition, the participants similarly assessed the ease at which they could find the two arrow designs among the other markings. Thus, it appears that the modified design performed as well as the current design. Researchers recommend implementing the modified wrong-way arrow design as existing wrong-way arrows are replaced.

Acknowledgments

This project was conducted in cooperation with TxDOT and the Federal Highway Administration. The authors thank Michael Chacon, Cary Choate, Ted Clay, Rick Cortez, Brian Fariello, John Gaynor, John Gianotti, Martin Gonzalez, Darrin Jenson, Omar Madrid, Wade Odell, and Roy Parikh of TxDOT who provided direction throughout the course of this research. The authors recognize Mack Cowan, Margaret Cunningham, and the many other DPS staff that assisted with the alcohol dosing and BAC measurements. The authors also acknowledge the contributions of Maury Dennis, Laura Higgins, Bari Kotwal, Albert Reeder, Sandra Stone, Nada Trout, Troy Walden, and Christine Yager of TTI, and the many other TTI staff that assisted in the design and conduct of this research.

Author Contributions

The authors confirm contribution to the paper as follows: study conception and design: MF, JM; data collection: JM; analysis and interpretation of results: MF, JM; draft manuscript preparation: MF, JM. All authors reviewed the results and approved the final version of the manuscript.

References

1. Finley, M. D., J. D. Miles, and E. S. Park. Closed-Course Study to Examine the Effect of Alcohol Impairment on a Driver's Ability to Identify and Read Signs. *Transportation Research Record: Journal of the Transportation Research Board*, 2017. 2660: 86–93.
2. Finley, M. D., S. P. Venglar, V. Iragavarapu, J. D. Miles, E. S. Park, S. A. Cooner, and S. E. Ranft. *Assessment of the Effectiveness of Wrong-way Driving Countermeasures and Mitigation Methods*. Report FHWA/TX-15/0-6769-1. Texas A&M Transportation Institute, College Station, TX, 2014. http://tti.tamu.edu/documents/0-6769-1.pdf. Accessed November 14, 2017.
3. *Highway Special Investigation Report: Wrong Way Driving*. Report NTSB/SIR-12/01. National Transportation Safety Board, Washington, D.C., 2012.
4. Tamburri, T. N., and D. J. Theobald. *Wrong-way Driving (Phase II)*. Traffic Department, Division of Highways, Department of Public Works, State of California, 1965.
5. Vaswani, N. *Further Reduction in Incidences of Wrong-way Driving*. Virginia Highway and Transportation Research Council, Charlottesville, SC, 1977.
6. Lathrop, S., T. B. Dick, and K. B. Nolte. Fatal Wrong-way Collisions on New Mexico's Interstate Highways, 1990–2004. *Journal of Forensic Sciences*, Vol. 55, No. 2, 2010, pp. 432–437.
7. Copelan, J. *Prevention of Wrong-way Accidents on Freeways*. Report No. FHWA/CA-TE-89-2. California

Department of Transportation, Traffic Operations Division, Sacramento, CA, 1989.

8. Scifres, P., and R. Loutzenheiser. *Wrong-way Movements on Divided Highways*. Purdue University Joint Highway Research Project No. JHRP-13-75, West Lafayette, Ind., 1975.
9. Cooner, S. A., A. S. Cothron, and S. E. Ranft. *Countermeasures for Wrong-way Movement on Freeways: Overview of Project Activities and Findings*. Report No. FHWA/TX-04/0-4128-1. Texas &M Transportation Institute, College Station, 2004. http://tti.tamu.edu/documents/4128-1.pdf. Accessed November 14, 2017.
10. Moler, S. Stop. You're going the Wrong-Way!. *Public Roads*, Vol. 66, No. 2, 2002, pp. 24–29. http://www.fhwa.dot.gov/publications/publicroads/02sep/06.cfm. Accessed November 14, 2017.
11. *Wrong-way Crashes in North Carolina Update*. Traffic Safety Unit, Transportation Mobility and Safety Division, North Carolina Department of Transportation, Raleigh, NC.
12. *Statewide Wrong-way Crash Study Final Report*. Florida Department of Transportation Central Office, Tallahassee, TN, 2015.
13. Zhou, H., J. Zhao, and R. Fries. *Investigation of Contributing Factors Regarding Wrong-way Driving on Freeways*. Report FHWA-ICT-12-010. Illinois Center for Transportation, Urbana, IL, 2012.
14. Baratian-Ghorghi, F., H. Zhou, and J. Shaw. Overview of Wrong-Way Driving Fatal Crashes in the United States. *ITE Journal*, Vol. 84, No. 8, 2014, pp. 41–47.
15. *Manual on Uniform Traffic Control Devices*. 2009 Edition with Revision Numbers 1 and 2. Federal Highway Administration, U.S. Department of Transportation, Washington, D.C., 2012. http://mutcd.fhwa.dot.gov/. Accessed November 14, 2017.
16. *Typical Standard Freeway Pavement Markings with Raised Pavement Markers FPM(1)-12*. Texas Department of Transportation, Austin, TX, 2012.

The Standing Committee on Traffic Control Devices (AHB50) peer-reviewed this paper (18-03121).

Article

TRR
JOURNAL OF THE TRANSPORTATION RESEARCH BOARD

Transportation Research Record
2018, Vol. 2672(21) 51–62

DOI: 10.1177/0361198118774235
journals.sagepub.com/home/trr
SAGE

Field Evaluation of the Dilemma Zone Protection System at Suburban Intersections

Sung Yoon Park[1], Chien-Lun Lan[1], Ranteg S. Rao[1], and Gang-Len Chang[1]

Abstract

Despite the fact that both traffic researchers and highway agencies have devoted considerable efforts over the past few decades to improving intersection safety, development of effective strategies to contend with this vital issue remains a challenging task. This research presents the field evaluation results of a Dilemma Zone Protection System (DZPS) implemented at two hazardous intersections in Maryland, U.S.A. The deployed DZPS can offer both proactive and reactive protections to drivers approaching a signalized intersection during a yellow phase. Field evaluations conducted at two intersections with DZPS deployed confirmed a 100% detection rate for red-light-running vehicles, timely activation of the all-red extension to prevent right-angle crashes, and effectively discouraging drivers from taking aggressive "pass" decisions during the yellow or all-red phases. The potential extension of the DZPS for speed harmonization on arterial traffic flows is also discussed.

Dilemma zones, as reported in the literature, are defined in two different ways. Gazis, Herman, and Maradudin define the dilemma zone as an area in which a vehicle approaching the intersection during the yellow phase can neither safely clear the intersection nor stop comfortably at the stop line (the GHM model or "Type-I Dilemma Zone) (*1*). Parsonson et al. define the dilemma zone, or "indecision zone" or Type-II Dilemma Zone, as a roadway segment before the stop line within which the probability of a driver choosing to stop when encountering a yellow phase drops from 90% to 10% (*2*).

Among a large body of studies, one of the popular methods of avoiding intersection incidents caused by the presence of dilemma zones is to provide a warning message in advance for drivers to reduce their approaching speeds. For example, Zimmerman et al. proposed a detection, control, and warning system (DCWS) to prevent drivers from being trapped in the dilemma zone (*3*). McCoy and Pesti concluded from a field study that an advance warning sign could reduce the number of vehicles trapped in dilemma zone (*4*). Some other studies also report that implementation of an advance warning sign can reduce red-light-running vehicles (ranging from 23% to 40%) (*5*, *6*) and the number of collisions (*7*).

Aside from providing warning messages, altering traffic signal timings to either extend or terminate the green phase to keep vehicles away from the dilemma zone is another proactive protection strategy (*8–13*). The detection–control system (D-CS) developed by the Texas Transportation Institute can dynamically identify an approaching vehicle's dilemma zone in advance and find the optimal duration of a green phase to minimize both the number of vehicles in the dilemma zone and the delays to vehicles in the other phases (*8*). Zegeer conducted a before and after evaluation on a spatially-designed green-extension system (GES) at three intersections in Kentucky. The field performance results showed a reduction of 54% in total crashes, and 75% in rear-end collisions, after deploying the GES (*9*).

Along the same line, Tarko et al. proposed a probabilistic approach to select the optimal duration of green extension to minimize the likelihood function of a GHM-model dilemma zone (*10*). Microprocessor-optimized vehicle actuation (MOVA) has been reported to reduce the number of dilemma zone crashes involving injuries by 30% (*11*). The incident reduction module in LHOVRA, a system developed in Sweden (*12*), was designed to reduce the number of vehicles trapped in the dilemma zone. A self-organized signal (SOS) system has been used to minimize the risks from the dilemma zone and delay on other

[1]Department of Civil and Environmental Engineering, University of Maryland, College Park, MD

Corresponding Author:
Address correspondence to Sung Yoon Park: alza102@gmail.com

approaches (*13*). New reactive protection measures, on the other hand, have been developed to prevent right-angle crashes when drivers fail to stop safely at a stop line. An example of such designs is the all-red extension, which can provide an extended all-red phase to accommodate vehicles failing to perform a stop at the intersection (*14*, *15*). Evaluation of the effectiveness of such an all-red extension method can also be found in the literature (*16–18*).

Despite increasing efforts by the traffic community on intersection safety, field evaluation of the effectiveness of a deployed DZPS remains an essential task. This study provides some insights into the resulting safety impacts by analyzing the before and after field-observed measure of effectiveness (MOE). The comparisons, based on the data collected over short- and long-term periods, are focused on the following critical issues:

- Can DZPS capture all red-light-running vehicles and provide all-red extensions in a timely manner to protect side-street traffic?
- Can DZPS affect the decision of approaching drivers? Does it encourage drivers to make a decision when encountering a yellow phase?
- Can DZPS effectively reduce the length of dilemma zones, which are dynamic in nature and vary with the approaching speed of drivers?

Design of DZPS

The DZPS deployed for the field evaluation was developed in our previous study (*19*), which has the following key features (Figure 1):

- *Long-range microwave detectors* for tracking the speed evolution of each vehicle within the detection zone. The detection zone covers the roadway segment from the stop line to the upper boundary of identified dilemma zones.
- *In-cabinet computer* for receiving traffic data from the detector(s), executing the embedded decision modules to determine the corresponding action plans, and sending all-red extension instructions to the traffic signal controller.
- *Signal controller* to provide signal phase and timing (SPaT) data to the in-cabinet computer and to receive the instructions for the all-red extension.

When the signal is in a yellow phase, the developed DZPS identifies vehicles approaching in the monitoring zone with their speeds and locations and calculates the probability of whether or not a detected vehicle will take the "pass" decision during the yellow and all-red phases. In appropriate cases, the system will extend the all-red phase with an additional clearance duration to prevent potential right-angled collisions.

Deployment of DZPS

For this study, DZPS was deployed at two high-speed suburban intersections in Maryland, U.S.A. Field evaluations were conducted at both locations; data was collected about 1 and 9 months after the activation of the deployed DZPS. Figure 2 shows the key geometric features of the two intersections and the configuration of the detection sensors at each.

Location Overview

The first location, the intersection of US 40 and Western Maryland Parkway (MD 910C) in Washington County, is a T-intersection consisting of two lanes on US 40 eastbound and westbound with a speed limit of 55 mph. Its

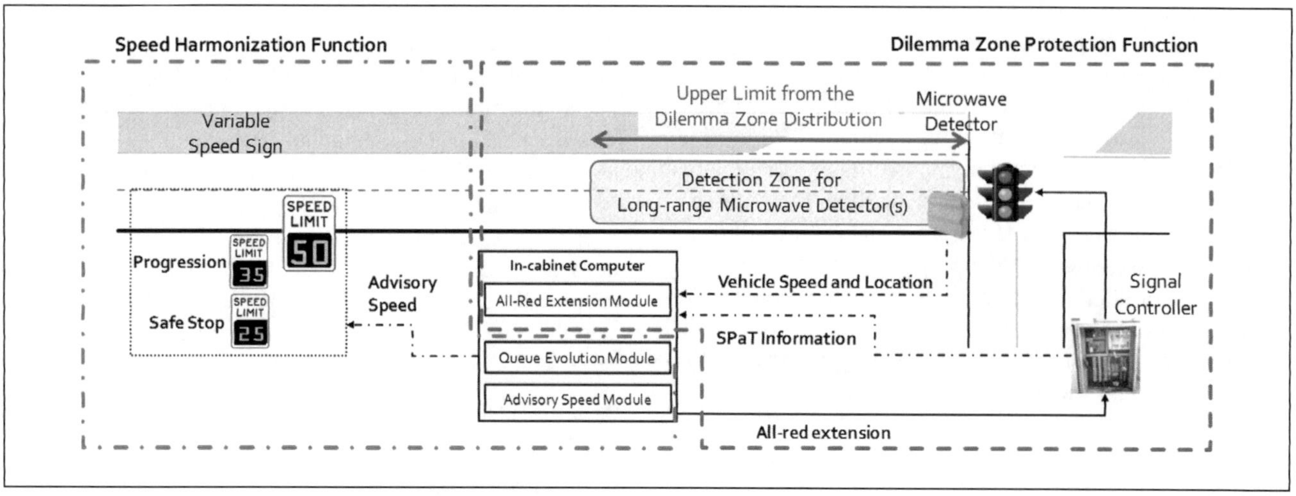

Figure 1. Design of the proposed intelligent intersection control system (DZPS and Speed Harmonization Function).

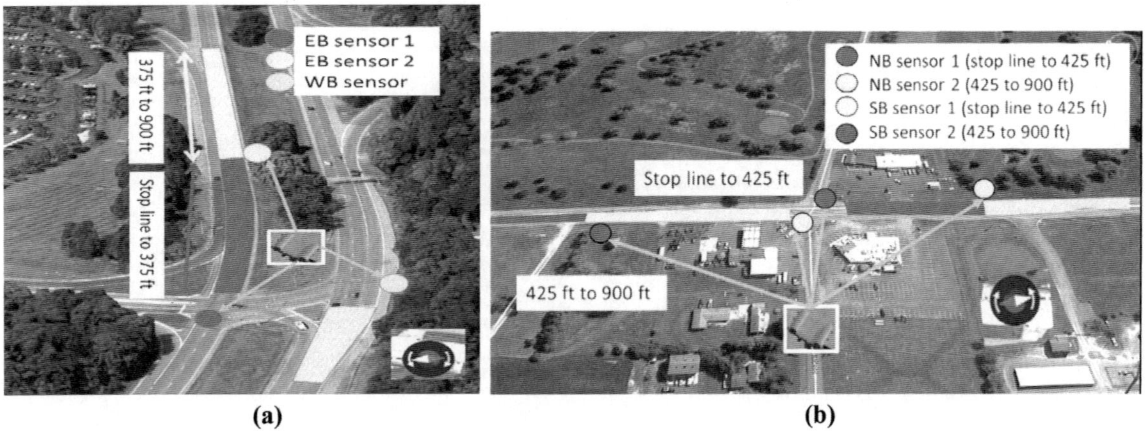

Figure 2. Deployment of DZPS at two intersections: (*a*) US 40 and Western Maryland Parkway (MD 910c); (*b*) MD 213 and Locust Point Road/Williams Street.

side street also has two lanes, one for left turns and the other for right turns only. This intersection experienced 15 crashes from 2011 to 2013; seven of which were right-angle crashes, and three cases resulted in personal injury. All the right-angle crashes involved the eastbound traffic (referred to as "US 40").

The second deployment site, the intersection of MD 213 and Locust Point Road/Williams Street in Cecil County, is a four-leg intersection with one lane in each approach and a speed limit of 55 mph. This intersection experienced six crashes from 2011 to 2013; four were right-angle crashes and one of them resulted in a fatality (referred to as "MD 213").

Field Data Collection

Field data was collected before deployment of DZPS using five camcorders located at 200, 300, 500, 700, and 900 ft, respectively, from the stop line to measure the speed of each approaching vehicle; one additional camcorder was used to collect the signal timings. After the deployment of DZPS, all data needed for before and after comparison was collected from both the deployed sensors and camcorders.

Performance evaluation, focusing on the effects of the deployed DZPS on traffic flow characteristics and intersection safety improvement, is based on the following measures of effectiveness (MOEs):

- MOE-1: Comparison of the before and after distribution of the dilemma zones, varying mainly with each individual vehicle's approaching speed and acceleration/deceleration rates;
- MOE-2: Before and after comparison of drivers' responses to the yellow phase at different approaching speeds and distances to the intersection stop line;
- MOE-3: Before and after comparison of traffic flow characteristics, including the speed distributions of approaching vehicles at different distances from the intersection stop line, and their average acceleration/deceleration rates;
- MOE-4: The detection rate of DZPS with respect to red-light-running vehicles and the false-alarm rate.

Impacts on the Traffic Flow Characteristics from the Deployed DZPS

Figure 3 shows the spatial distribution of the speeds derived from the field data. As reflected in their patterns, the deployed DZPS does not eliminate all speeding vehicles (see Table 1). After the deployment, however, the average traffic flow speeds at different distance-to-stop lines decreased significantly, compared with the data observed before the deployment. For example, at the intersection of US 40, the average speed reduced from 49.7 mph to 44.6 mph at the distance of 900 ft from the stop line, and from 40 mph to 34.9 mph at 200 ft (Figure 3*a*). As shown in Figure 3*b*, the same decreasing patterns were also observed at the intersection of MD 213. As shown at the bottom of these figures, *t*-tests also indicate the differences in speeds before and after the deployment. Hence, one can conclude that deployment of the DZPS sensors indeed has some effect on drivers approaching those intersections. Most drivers are likely to reduce their speeds, except for aggressive drivers approaching the intersections at speeds far exceeding the posted speed limit.

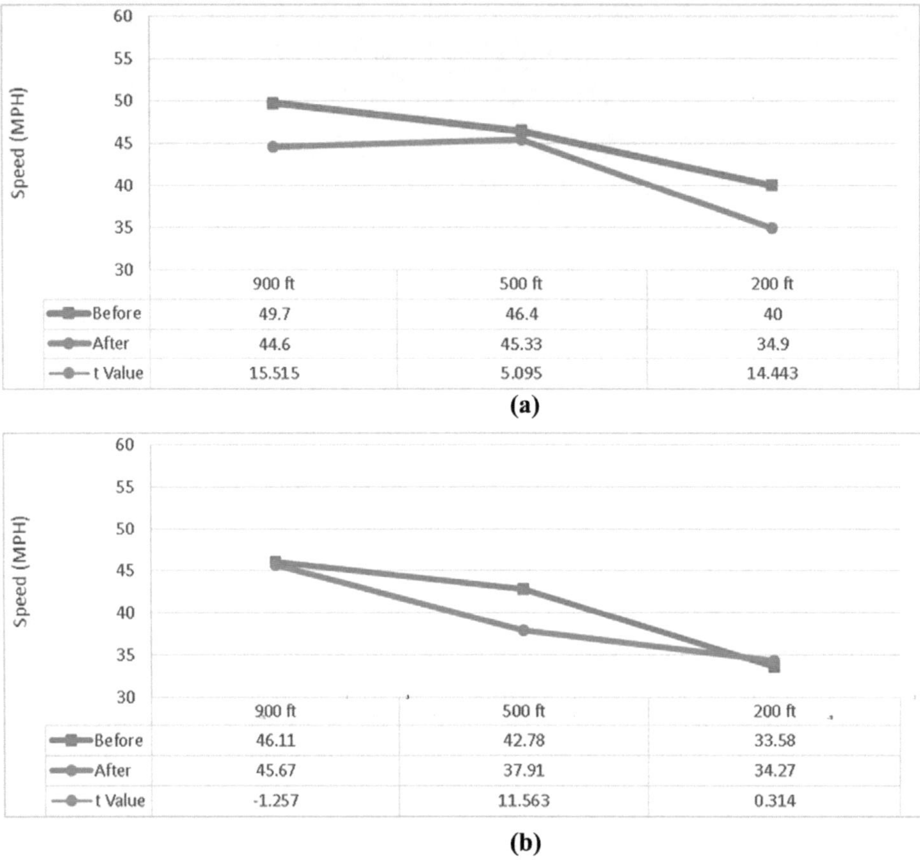

Figure 3. Comparison of the spatial distribution of speeds before and after the deployment of DZPS at two intersections: (*a*) US 40; (*b*) MD 213.

Table 1. Frequency Distributions by Speed Bin for Approaching Vehicles before and after DZPS Deployment at 900 ft from Intersections

	US40 intersection				MD 213 intersection			
	Before		After		Before		After	
Speed (mph)	Frequency	Percentage	Frequency	Percentage	Frequency	Percentage	Frequency	Percentage
75 +	14	1	0	0	N/A	N/A	N/A	N/A
70–75	36	3	3	0	3	0	0	0
65–70	58	5	6	0	3	0	2	0
60–65	92	7	94	3	8	1	0	0
55–60*	160	13	375	13	37	6	7	1
50–55	189	15	850	29	113	18	32	5
45–50	206	17	951	32	177	29	115	17
40–45	236	19	432	15	147	24	254	38
35–40	153	12	166	6	69	11	182	27
30–35	87	6	66	2	58	9	77	12
Over the speed limit (total)	360 (1231)	29	478 (2943)	16	51 (615)	8	9 (669)	1

*Speed limit for both US 40 and MD 213.

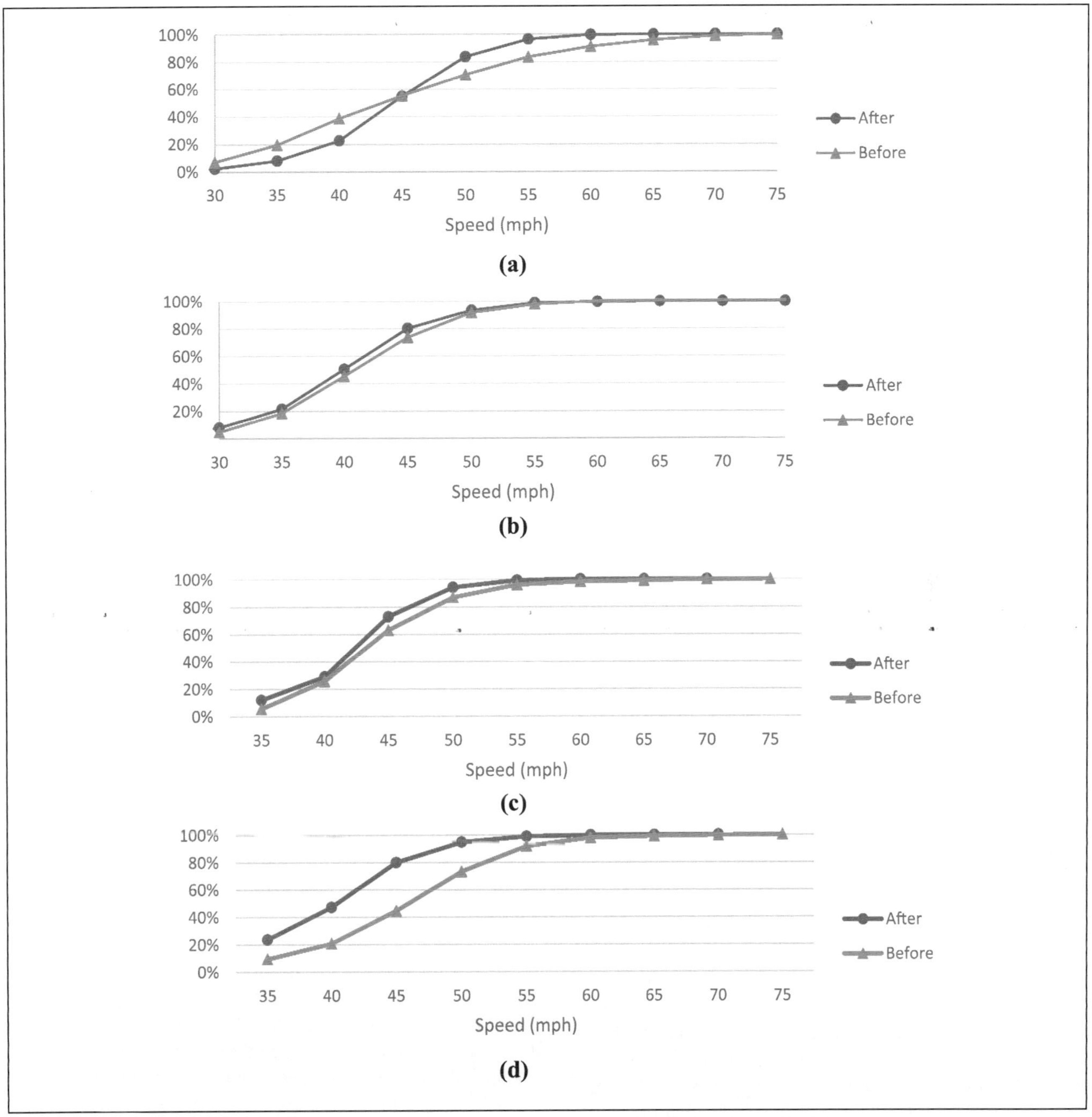

Figure 4. Cumulative distribution of approaching speeds before and after DZPS deployment: (*a*) at a distance of 900 ft (US 40); (*b*) at a distance of 500 ft (US 40); (*c*) at a distance of 900 ft (MD 213); (*d*) at a distance of 500 ft (MD 213).

Figure 4 shows the cumulative speed distributions before and after the deployments at different locations, and Table 1 summarizes the percentage of vehicles within each speed bin (every five mph range). Note that the speed limit at the location for the DZPS deployment is 55 mph. As shown in Table 1, the percentage of high-speed vehicles has been reduced from 29% to 16%, a reduction of 13% for US 40 and from 8% to 1% for MD 213. These statistics further highlight the effectiveness of the deployed DZPS. Furthermore, the cumulative distributions of drivers at different speeds at the locations of 900 ft and 500 ft clearly show that the deployed system can increase the percentage of drivers approaching the intersection at speeds below or around the posted speed limit.

The deployed system did not eliminate all aggressive drivers (i.e., over the speed limit) observed at both study sites, however. This justifies the need to implement some reactive control strategies (e.g., the all-red extension) to

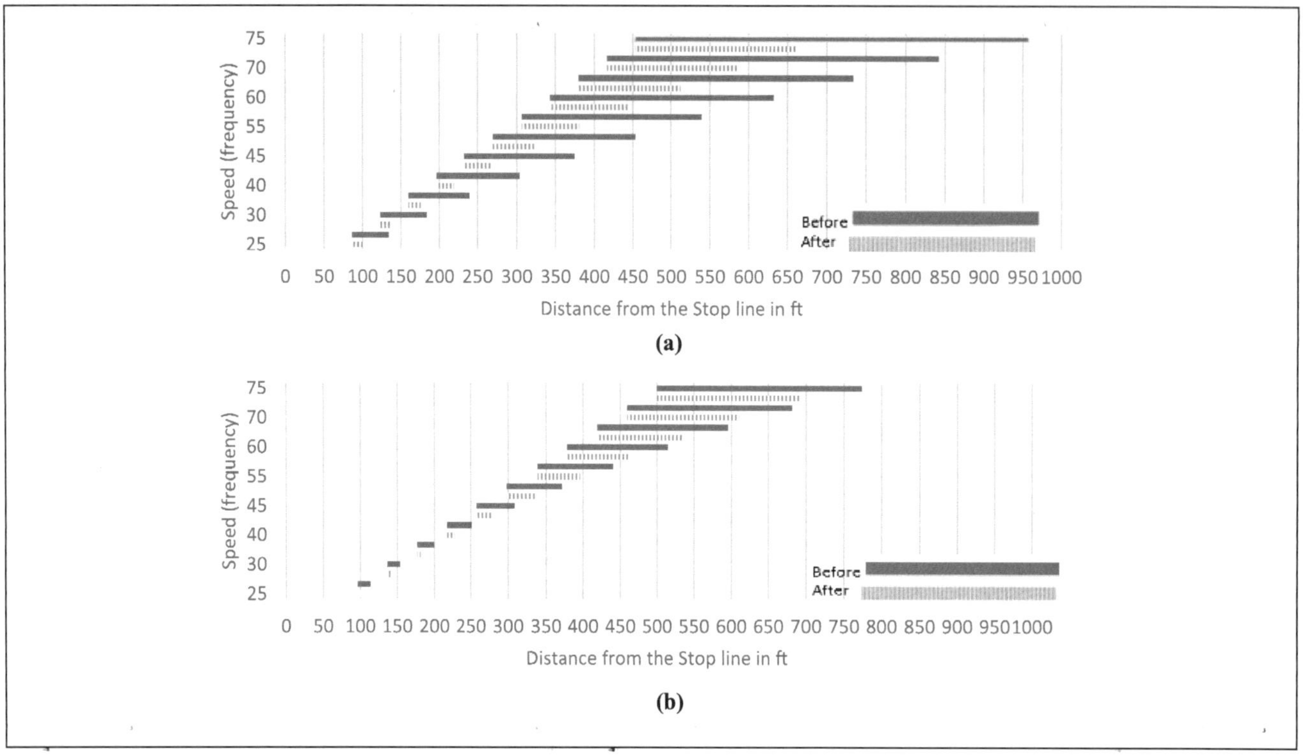

Figure 5. Before and after distribution of dilemma zones: (*a*) US40; (*b*) MD 213.

prevent those drivers from colliding with vehicles entering the intersection from side streets.

Impacts on Deceleration Rates and the Distribution of Dilemma Zones

The average acceleration/deceleration rates of approaching vehicles, either choosing the "stop" action during the yellow phase or the "pass" decision over the all-red phase after deployment of DZPS, were calculated from the recorded videos and the sensor log files. It is noticeable that the average deceleration rate was changed from $-7.28\ ft/s^2$ to $-11.27\ ft/s^2$ after the DZPS deployment at US 40, and from $-9.79\ ft/s^2$ to $-10.69\ ft/s^2$ at the intersection of MD 213. An increase in the average deceleration rate implies that drivers are more willing to take the "stop" decision at the onset of a yellow phase after seeing the deployed DZPS devices.

Based on the newly calculated acceleration/deceleration rates, one can then compute the distribution of dilemma zones after the system deployment (Figure 5), based on the equation shown in the literature (*1*), that is:

$$x_{dz} = v_0\delta_2 + \frac{v_0{}^2}{2a_2{}^*} - v_0\tau + (w + L) - \frac{1}{2}a_1{}^*(\tau - \delta_1)^2 \quad (1)$$

where:

τ : Duration of yellow interval (second);

δ_1 :Reaction-time lag of the driver–vehicle complex (second);

δ_2: Decision-making time of a driver (second);

v_0: Approaching speed of vehicles (ft/s);

$a_1{}^*$:Maximum acceleration rate of approaching vehicles (ft/s^2);

$a_2{}^*$: Maximum deceleration rate of approaching vehicles (ft/s^2);

w: Intersection width (ft); and

L: Average vehicle length (ft).

Clearly, the length of the dilemma zone for each speed bin has been reduced, compared with before the deployment. For example, Figure 5*a* shows that the upper limit of the dilemma zone was reduced from 960 to 670 ft for vehicles traveling at 75 mph, and from 840 to 580 ft for vehicles traveling at 70 mph.

Table 2 summarizes the decisions of drivers during a yellow phase at the two intersections. The percentage of drivers taking the "pass" action at the onset of a yellow phase decreased for those approaching the intersection at a moderate speed (i.e., 45–55 mph for the speed limit of 55 mph). For example, for vehicles traveling at 45 to 55 mph at a distance of between 300 and 400 ft from the stop line at the onset of the yellow phase, at data collection before deployment of DZPS on US 40 about 50% of drivers chose to pass the intersection, but after deployment it decreased to 43% (decrease from 50% to 18% on MD 213).

Table 2. Before and After Comparison of Drivers Taking the "Pass" Decision at the Intersection after One Month of DZPS Deployment

Speed of vehicles at the onset of a yellow phase	Distance of vehicles from the stop line at the onset of a yellow phase									
	0–100 ft		100–200 ft		200–300 ft		300–400 ft		400 + ft	
	Before	After	Before	After	Before	After	Before	After	Before	After
(*a*) US 40 Intersection										
45–55 mph	100% (78)	100% (24)	100% (100)	94% (32)	74% (73)	59% (41)	50% (24)	43% (40)	20% (5)	5% (59)
55 + mph	100% (9)	100% (7)	100% (20)	100% (9)	88% (47)	91% (22)	50% (16)	54% (13)	10% (20)	9% (44)
(*b*) MD 213 Intersection										
45–55 mph	100% (78)	100% (13)	100% (100)	100% (6)	74% (73)	43% (21)	50% (24)	18% (28)	20% (5)	3% (61)
55 + mph	100% (9)	100% (2)	100% (20)	100% (5)	88% (47)	100% (6)	50% (16)	44% (21)	10% (20)	19% (31)

Note: The number in parenthesis denotes the sample size.

The reduction in "pass" decision among the drivers varied between 6% and 15% on US 40 (17% and 32% on MD 213), depending on distance from the intersection when encountering the yellow phase.

For drivers approaching the intersection at a speed above the speed limit, the presence of DZPS seems to have no influence on their decision during the yellow phase. The percentage of high-speed vehicles taking the "pass" decision during a yellow phase remained relatively stable at data collection before and after deployment, demonstrating much less change than those approaching the intersection at a moderate speed under the same decision scenarios (Figure 6). Hence, since the resulting patterns at both intersections are consistent, one can conclude that the presence of the DZPS's roadside sensors may discourage most non-aggressive drivers, traveling at less than or around the speed limit, from taking the "pass" decision during a yellow phase (i.e., from 20% to 5% at more than 400 ft), but not for those aggressive drivers (i.e., from 10% to 9% at more than 400 ft).

A similar analysis was conducted after 9 months of the system deployment. Table 3 shows the decisions of those drivers during the yellow phase when they approached the intersection (a) before, (b) 1 month, and (c) 9 months after deployment of the DZPS at the two intersections. As shown in the table, no discrepancy is discernible for moderate drivers (i.e., 45–55 mph), but the percentage of aggressive drivers (i.e., >55mph), who choose to take the "pass" decision now exhibits a distinct reduction when encountering a yellow phase at the distance of more than 200 ft from the stop line.

Performance Evaluation

The performance evaluation of DZPS with respect to the detection rate of red-light-running vehicles and timely execution of all-red extensions includes the following steps:

- Identify any red-light-running vehicles from the recorded videos;
- Identify each all-red phase from the recorded videos;
- Determine whether or not the all-red extension was executed for each identified red-light-running vehicle; and
- Calculate the detection rates, all-red extension rates, and false-alarm rates.

Table 4 shows the results of the above performance evaluation. The selected MOEs for the evaluation are: (a) the extension-call rate which shows the number of all-red extensions over the total number of cycles; (b) the detection rate which indicates the number of detected red-light-running vehicles over the total number of red-light-running vehicles; and (c) the false-alarm rate which denotes the number of all-red extensions without red-light-running vehicles over the total number of cycles.

As shown in Table 4, the deployed DZPS successfully detected all red-light-running vehicles, and executed all-red extensions in time for those aggressive drivers. However, the 100% detection rate was accomplished at the cost of false-alarm rates of 30% and 16%, respectively, at the two intersections. This is due to the fact that the DZPS is designed and deployed to ensure the safety of the intersection, and thus the detection of red-light-running vehicles should be maximized and the false-alarm rate has been compromised.

The optimal tradeoff between the detection and false-alarm rates can be investigated theoretically. Abadi et al. found that all-red extension calls between different thresholds of system accuracy had little effect on traffic delays (*18*). Considering the potential cost of a red-light-

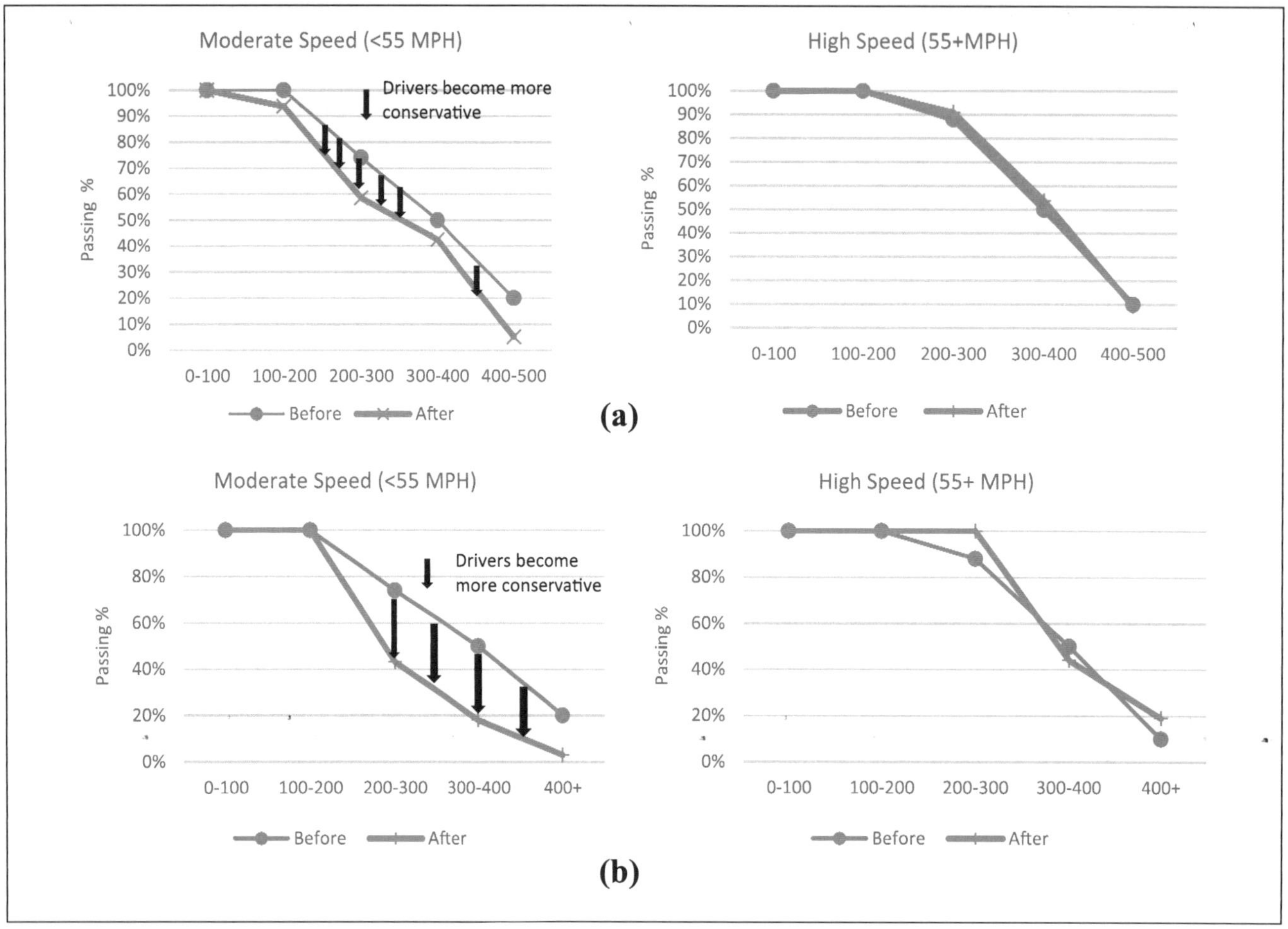

Figure 6. Before and after comparison of drivers taking the "pass" decision during the yellow phase under different approaching speeds (speed limit 55 mph): (*a*) US 40; (*b*) MD 213.

running vehicle causing a right-angle crash compared with unnecessary delays to traffic flow due to some false alarms, the detection rate of 100% was set as the foremost objective of the deployed DZPS.

System Design Extension with Speed Harmonization Function

The deployed DZPS can be extended to execute a proactive speed harmonization function with minimal hardware enhancements (see Figure 1). Figure 7 illustrates the data flows and computation modules for the intersection speed harmonization function. The long-range microwave detector (from DZPS) can provide each individual vehicle's location and speed within the detection area. The discharge and arrival flow rates at the intersection can also be derived. The signal phasing and timing data, provided by the signal controller, are also sent to the computation modules (Figure 7). When the intersection is operated under actuated signal control, a supplemental module to estimate the green termination timing will be executed to compute the remaining phase duration.

The proposed speed harmonization function shall have three essential computation modules:

- *Estimate the number of vehicles within the control boundaries*: This module functions to estimate the number of vehicles from the intersection stop line to the advisory speed sign. The proposed function computes the traffic density within the detection zone, using vehicle counts directly available from the system's sensor. The computed density is then used to estimate the number of vehicles between the stop line and the advisory speed sign, which may be located beyond the dilemma zone.
- *Predict the time-varying stop queues*: This function is to predict the queue length using traffic information from both system sensors and signal states from the traffic controller.
- *Estimate the probability of an observed vehicle taking the "pass" action during a yellow phase*: This

Table 3. Before and After Comparison of Drivers Taking the "Pass" Decision

Speed at the onset of a yellow phase	Location of vehicles from the stop line at the onset of a yellow phase														
	0–100 ft			100–200 ft			200–300 ft			300–400 ft			400 + ft		
	Before	1 month	9 month	Before	1 month	9 month	Before	1 month	9 month	Before	1 month	9 month	Before	1 month	9 month
(*a*) US 40 Intersection															
45–55 mph	100%	100%	100%	100%	94%	96%	74%	59%	83%	50%	43%	50%	20%	5%	0%
55 + mph	100%	100%	100%	100%	100%	100%	88%	91%	50%	50%	54%	43%	10%	9%	8%
(*b*) MD 213 Intersection															
45–55 mph	100%	100%	100%	100%	94%	100%	74%	43%	63%	50%	18%	30%	20%	3%	0%
55 + mph	100%	100%	100%	100%	100%	100%	88%	88%	70%	50%	44%	33%	10%	19%	5%

Note: Speed limit of 55 mph at both intersections. Measurements taken before, 1 month after, and 9 months after deployment of DZPS.

Table 4. Evaluation of Detection Performance of the DZPS

MOE	Field operation* (US 40)	Field operation** (MD 213)
Red-light-running rate (RLR) (RLR / cycle)	1.6%	1.6%
Extension-call rate (extension call / cycle)	31.7%	17.6%
Detection rate (protected RLR)	100.0%	100.0%
False-alarm rate (false alarm / cycle)	30.0%	16.0%

* Five red-light-running vehicles during the data collection (312 cycles)
** Seven red-light-running vehicles during the data collection (441 cycles)

module, the core of the DZPS, can track each individual vehicle's trajectory within the detection area, and estimate a vehicle's pass/stop probability during a yellow phase to determine the activation of an all-red extension (*18*).

The computation module will determine the advisory speeds for either promoting a safe stop or encouraging progression. Vehicles arriving at the intersection at the end of red phases may be able to progress through the intersection by delaying their arrivals until all stop queues are cleared. An advisory speed, calculated from the queue evolution and remaining phase duration, will be posted to encourage drivers to stay in the progression band. Vehicles arriving at the end of a green or at a red phase, on the contrary, will need to stop safely at the end of the queue. An advisory speed, considering both queue length and available stopping distances, can inform drivers of the phase changes and reduce the need for hard braking.

Evaluation of the Speed Harmonization Function using a Simulation Platform

This study has designed a simulation platform to evaluate the effectiveness of the proposed speed harmonization function. The key computation modules, system functions, and driver responses are developed with C# programming language; whereas the simulation kernel is implemented with VISSIM. The simulation platform has been carefully calibrated with field data and adopted the concept from Park et al. (*19*) to approximate the effect of advisory speeds on traffic flow speeds.

As protecting vehicles from crashes is the top priority of the proposed system, this study has tuned the all-red extension module to offer a 100% detection rate, which is expected to result in a relatively high false-alarm rate

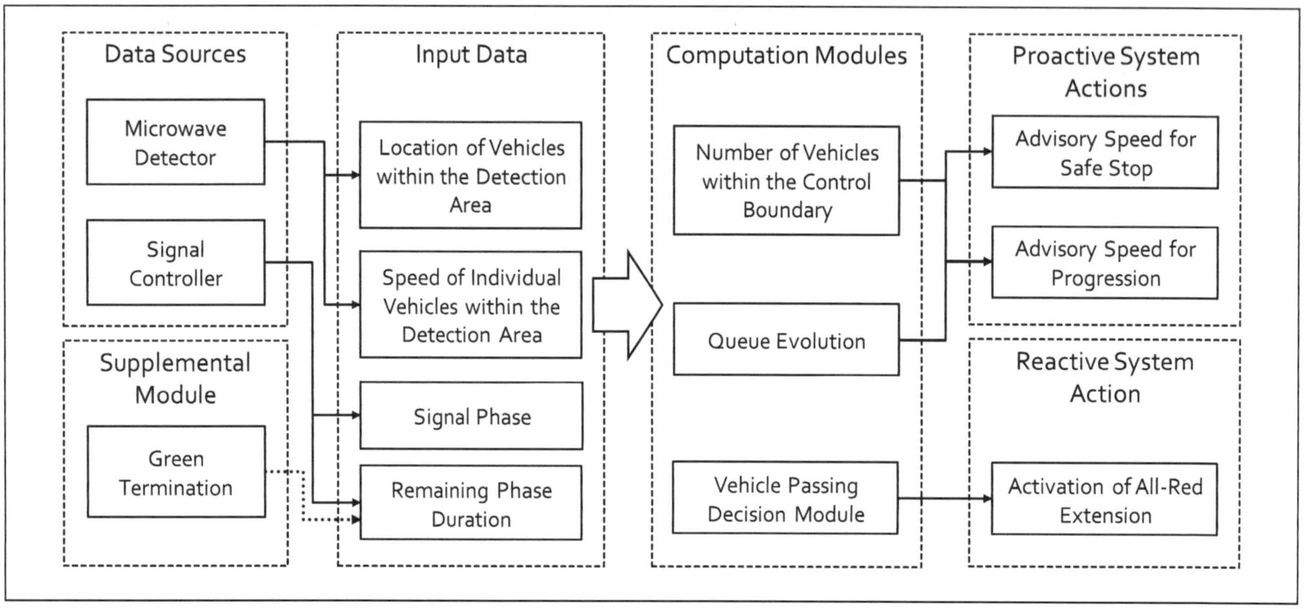

Figure 7. Data flows between key computation modules of the DZPS extended functions.

Table 5. System Performance for Different Test Scenarios with Actuated Signal Control

Scenarios MOEs		Base case	All-red extension only	All-red extension and speed harmonization function[d] with different compliance rates 100%	50%	25%
Safety MOEs	Average number of hard-braking vehicles per 100 cycles[a]	76	77	18	21	26
	Total number of vehicles in dilemma zone	792	770	525	596	622
	The average number of red-light-running vehicles per cycle (RLR / 100 cycle)[b]	9.4	9.7	6.4	8.1	8.7
	Detection rate for red-light-running vehicles	na	100%	100%	100%	100%
Mobility MOEs	Average number of stops per cycle	6.4	6	4.5	5.1	6.89
	Average stopped delay for all vehicles per cycle (seconds)	76	77	54	59	63
	Average fuel consumption per cycle (gallon)	0.24	0.22	0.14	0.15	0.17
	False-alarm rate for the all-red extension[c]	na	16%	4%	7%	7%

[a]Total number of hard-braking vehicles divided by the total number of cycles (deceleration rate >10 ft/s^2).
[b]Total number of red-light-running vehicles divided by the total number of cycles.
[c]Total number of false all-red extensions divided by the total number of cycles.
[d]Advisory speed: limited to the range of 30 to 50 mph with an increment of 5 mph.

(i.e., 16%). Table 5 shows that introduction of the speed harmonization function can reduce the average number of hard-braking vehicles per 100 cycles from 76 to 18 under a driver compliance rate of 100%. One may also note from the simulation results (even with low driver compliance rates) that it is still effective in reducing potential rear-end crashes (i.e., using hard-braking vehicles as a surrogate variable). As for the number of vehicles trapped within the dilemma zone, the result shows a reduction of 33%.

The proposed system is also effective in improving traffic mobility. As shown in Table 5, the average number of stopped vehicles per cycle is reduced from 6.4 to 4.5, and the average stop delay per cycle is reduced from 76 to 54 seconds. With the improved mobility, the fuel consumption shows a significant reduction from 0.24 to 0.14 gallon per cycle. One may also notice that with the advisory speed the false-alarm rate can be reduced from 16% to 4%, which can improve the traffic mobility by reducing excessive all-red extension calls. Based on the results of our preliminary simulation experiments, one can conclude that the proposed speed harmonization extension has the potential to perform effectively the following functions:

- The advisory speed sign, activated before the onset of a yellow phase, can effectively reduce the number of vehicles trapped in the dilemma zones and minimize potential rear-end crashes;
- The intersection's operational efficiency (e.g., the number of stops, stop delays, and fuel consumption) can be improved with the smoother traffic progression guided by the advisory speed function.

Conclusion

To confirm the effectiveness of the proposed DZPS, this study has collected before and after field data from two high-speed suburban intersections, and compared traffic characteristics at both locations. The collected data includes the approaching speed distributions, deceleration/acceleration rates, drivers' response to yellow phases, and detection rates of red-light-running vehicles. Our extensive data analyses confirm that the deployment of DZPS has (a) lowered approach speeds, (b) reduced the probability of drivers taking a "pass" decision during a yellow phase, and (c) reduced the length of the dilemma zone. It has also been observed from the field data that the system has achieved 100% detection rates for red-light-running vehicles.

To improve both the safety and mobility of traffic at high-speed intersections, this study has presented an extension to execute a speed harmonization function, using the existing DZPS hardware devices. The proposed system can operate either proactively or reactively to mitigate crashes and promote traffic progression by advising drivers to take appropriate speeds in real time. The extended use of the DZPS hardware and control components allows responsible agencies to best use available resources for concurrently improving the safety and operational efficiency. From the extensive experiments with a well-calibrated simulator, the proposed system has demonstrated its effectiveness in offering full protection against red-light-running vehicles and in reducing the potential for rear-end crashes. The simulation results also show the system contributes to fewer stops, reduced stop delays, and lower fuel consumption as the result of smooth traffic flows guided with the dynamic advisory speed signs.

Future research along the same lines should address the following issues: (a) evaluating the long-term safety improvement (i.e., number of right-angle crashes), (b) improving the traffic progression over several intersections (an arterial level), and (c) incorporating the presence of connected vehicles in the traffic flows under various penetration rates. In addition, the fact that information on traffic queue evolution near the stop line is available from the DZPS's long-range sensors, makes it possible to design an advance signal plan to satisfy concurrently the needs of safety improvement and delay minimization.

Acknowledgments

This study was sponsored by MDSHA (Maryland State Highway Administration). The authors are grateful to MDSHA and its staff for their continuing support, especially Allison Heart, Xiang Hua, Dilip Patel, Ruihua Tao, and Robert Deblase.

Author Contributions

The authors confirm contribution to the paper as follows: study conception and design: Sung Yoon Park, Chien-Lun Lan, and Gang-len Chang; data collection: Sung Yoon Park and Chien-Lun Lan; analysis and interpretation of results: Sung Yoon Park, Chien-Lun Lan, and Ranteg Rao; draft manuscript preparation: Sung Yoon Park. All authors reviewed the results and approved the final version of the manuscript.

References

1. Gazis, D., R. Herman, and A. Maradudin. The Problem of the Amber Signal Light in Traffic Flow. *Operations Research*, Vol. 8, 1960, pp. 112–132.
2. Parsonson, P.S., R.W. Rosevearem, and J. R. Thomas Jr. Small-Area Detection at Intersection Approaches. *Traffic Engineering*, Vol. 44, 1974, pp. 8–17.
3. Zimmerman, K., D. Tolani, R. Xu, T. Qian, and P. Huang. Detection, Control, and Warning System for Mitigating Dilemma Zone Problem. *Transportation Research Record: Journal of the Transportation Research Board*, 2014. 2298: 30–37.
4. McCoy, P. T., and G. Pesti. Dilemma Zone Protection on High-Speed Signalized Intersection Approaches: Advance Detection Versus Advance Warning Flashers and Advance Detection. Presented at 83rd Annual Meeting of the Transportation Research Board, Washington, D.C., 2004.
5. Bowman, B. L. *NCHRP Synthesis of Highway Practice 186: Supplemental Advance Warning Devices, A Synthesis of Highway Practice*. TRB, National Research Council, Washington, D.C., 1993.
6. Messer, C. J., S. R. Sunkari, H. A. Charara, and R. T. Parker. *Development of Advance Warning Systems for*

End-of-Green Phase at High Speed Traffic Signals. Publication FHWA/TX-04/0-4260-4. Texas Transportation Institute, Texas A & M University System, Austin, Texas, 2004.

7. Sayed, T., H. Vahidi, and F. Rodriguez. Advance Warning Flashers: Do They Improve Safety? *Transportation Research Record: Journal of the Transportation Research Board*, 1999. 1692: 30–38.
8. Bonneson, J. A., D. Middleton, K. Zimmerman, H. Charara, and M. Abbas. *Intelligent Detection-Control System for Rural Signalized Intersections.* Publication FHWA/TX-03/4022-2, Texas Transportation Institute, Austin, Texas, 2002.
9. Zegeer, C. V. *Effectiveness of Green-Extension Systems at High-speed Intersections.* Department of Transportation, Lexington, Kentucky, 1977.
10. Tarko, A., W. Li, and L. Laracuente. Probabilistic Approach to Controlling Dilemma Occurrence at Signalized Intersections. *Transportation Research Record: Journal of the Transportation Research Board*, 2006. 1973: 55–63.
11. Peirce, J. R., and P. J. Webb. *MOVA: The 20 Site Trial.* TRRL Research Report RR279, Great Britain, 1990.
12. Peterson, A. *LHOVRA a New Traffic Signal Control Strategy for Isolated Junctions.* Traffic Engineering and Control, Sweden, 1986.
13. Kronborg, P., F. Davidsson, and J. Edholm. *SOS–Self Optimizing Signal Control: Development and Field Trials of the SOS Algorithm for Self-Optimizing Signal Control at Isolated Intersections.* TFK Report 1997:2E, Transport Research Institute, Stockholm, Sweden, 1997.
14. Zhang, L., L. Wang, K. Zhou, and W. Zhang. Dynamic All-Red Extension at a Signalized Intersection: A Framework of Probabilistic Modeling and Performance Evaluation. *IEEE Transactions on Intelligent Transportation Systems*, Vol. 13, 2012, pp. 166–179.
15. Gates, T. *A Dynamic On-Demand All-Red Clearance Interval Extension Process for Stochastic Vehicle Arrivals at Signalized Intersection.* PhD dissertation. University of Wisconsin–Madison, 2007.
16. Hurwitz, D., M. G. Abadi, S. McCrea, S. Quayle, and P. Marnell. *Smart Red Clearance Extensions to Reduce Red-Light Running Crashes.* Publication FHWA-OR-RD-16-10, Oregon Department of Transport, Salem, Oregon, 2016.
17. Chang, G. L., M. L. Franz, Y. Liu, Y. C. Lu, and R. Tao. *Design and Evaluation of a Dynamic Dilemma Zone System for a High Speed Rural Intersection.* Report No. SP708B4J, Maryland Department of Transportation, Baltimore, Maryland, 2012.
18. Abadi, M. G., D. S. Hurwitz, P. Marnell, and S. Quayle, Evaluation of Red Clearance Extension Designs with Hardware-in-the-Loop Simulation, *Transportation Letters*, 2017. (In press).
19. Park, S., C. L. Lan, G. L. Chang, D. Tolani, and P. Huang. Design and Predeployment Assessment of an Integrated Intersection Dilemma Zone Protection System. *Journal of Transportation Engineering*, 2016. (In press).

The Standing Committee on Traffic Control Devices (AHB50) peer-reviewed this paper (18-03064).